SpringerBriefs in Operations Management

Series Editor

Suresh P. Sethi
School of Management
The University of Texas at Dallas
Richardson, TX, USA

More information about this series at http://www.springer.com/series/13082

Shuo Zeng • Moshe Dror

Formulating Principal-Agent Service Contracts for a Revenue Generating Unit

Springer

Shuo Zeng
Eller College of Management
University of Arizona
Tucson, AZ, USA

Moshe Dror
Eller College of Management
University of Arizona
Tucson, AZ, USA

SpringerBriefs in Operations Management
ISBN 978-3-319-18671-9 ISBN 978-3-319-18672-6 (eBook)
DOI 10.1007/978-3-319-18672-6

Library of Congress Control Number: 2015942238

Springer Cham Heidelberg New York Dordrecht London

Printed on acid-free paper

Springer International Publishing AG Switzerland is part of Springer Science+Business Media (www.springer.com)

Contents

List of Figures

List of Tables

Chapter 1
Introduction

With the ongoing technological advancements in manufacturing, health delivery systems, information technologies etc., numerous industrial entities become reliant on sophisticated product delivery systems for provision of revenue generating operations. For example, fuel-efficient aircraft engines are essential for airlines to provide affordable transportation services; mining companies operate large interdependent mining equipment units for extraction of hundreds truck loads of ores everyday; oil refineries construct groups of fractionating columns to produce various crude oil products; sophisticated flexible manufacturing systems enable manufacturing companies to machine different types of parts with high efficiency at low costs; data server arrays are the backbone of real-time electronic transaction systems operated by banks and credit card companies; advanced office printing and scanning equipment is indispensable for efficient information collection and dissemination in large companies, universities and government agencies.

There are common characteristics shared by the equipment units of sophisticated product delivery systems. First, the equipment units are mission-critical such that no revenue is generated when the equipment units fail. Second, these units are assumed to operate in a reliable mode with short downtimes relatively to their uptimes. Third, the units are usually of a specialized nature that requires expert maintenance/service providers. It is known for the owner of such systems to outsource the maintenance and repair of her equipment units to an independent supplier of specialized repair services. Therefore the main topic of this paper – the analysis of the contractual details that have to be addressed in the agreement between the system's owner and the supplier of maintenance and repair services.

In this paper we examine the contractual options between the owner (principal, she) of a revenue generating unit and a service provider (agent, he) in a framework of principal-agent economic model. Although our initial framing of the principal-agent problem follows Kim et al. (2010), our analysis is significantly different from Kim et al. (2010) and is much more extensive than their analysis. First, the

S. Zeng, M. Dror, *Formulating Principal-Agent Service Contracts for a Revenue Generating Unit*, SpringerBriefs in Operations Management,
DOI 10.1007/978-3-319-18672-6_1

agent is assumed to be risk-neutral or risk-averse in Kim et al. (2010) while our analysis includes risk-seeking agent also. Second, our analysis of the principal-agent contract covers the value of exogenous parameters exhaustively, while Kim et al.'s (2010) assumptions of a reliable equipment unit and negligible downtimes (compared to uptimes) require that the values of certain exogenous parameters fall into a narrow range. Finally and the most important, we derive explicit formulas for optimal principal-agent contract under any market and industry conditions without imposing any additional constraint, while Kim et al. (2010) is able to provide only one explicit formula when Service Time Target Constraint is binding.

In a counter-distinction to Kim et al. (2010) we model the principal-agent system of a risk-neutral principal with risk-neutral, risk-averse, or risk-seeking agent as a Markov process with an undetermined time horizon instead of a contract for a finite horizon normalized to 1. In addition, we replace Kim et al.'s (2010) representation of agent's risk as variance of his revenue stream with a piece wise linear function in a steady state probability of failure as a proxy for a measure of agent's revenue risk.

Our analysis is restricted to a single risk-neutral principal who owns one unit of revenue generating entity and a single agent.

Chapter 2
The Basic Principal-Agent

In a basic principal-agent setting, the principal contracts an agent to perform a service function and the agent chooses the level of his capacity (his 'effort') in response to the contract offer and subsequently its effect on the principal's revenue stream. We assume that the principal's equipment unit generates revenue at an expected rate of $r > 0$ \$ per unit of uptime. The unit runs for a random period of time before failing, and remains in the failed state until it is repaired. To address the recurring maintenance and equipment failures the principal contracts an agent who subsequently installs a repair capacity and repairs the principal's equipment when it fails. The contract structure considered is rather simple: the principal proposes to pay the agent $w > 0$ \$ per unit of time during the duration of the contract but the agent pays the principal $p > 0$ \$ per unit of time during the unit's failure duration. The agent's capacity decision is unobservable by the principal. Each party is presumed to choose the values that maximize his/her utilities. We assume that the parties are rational and each knows that the other is rational, etc. till infinitum. It includes their individual computational ability to anticipate (compute) the other's best response to any offer. Therefore, with some abuse of timing we presume that both, the contract offer and the service capacity decision, occur at the same time with full knowledge of the two parties.

In general, if the agent's action is observable and contractible, then the principal would contract directly on agent's service capacity that maximizes the principal's profit leaving zero surplus to the agent – enough to ensure agent's participation. Such a scenario is referred to as the *first-best solution* (Hölmstrom 1979). If the agent's action is unobservable and therefore uncontractible, then the agent's response may deviate from the one prescribed by the principal in the first-best solution, and the principal risks realizing lower profits. The likelihood and the degree of agent's deviation from the desired action is referred to as *moral hazard* (Luenberger 1995). When moral hazard is present the principal uses the available

S. Zeng, M. Dror, *Formulating Principal-Agent Service Contracts for a Revenue Generating Unit*, SpringerBriefs in Operations Management,
DOI 10.1007/978-3-319-18672-6_2

information about the agent's action to alleviate the moral hazard (Hölmstrom 1979) and proposes a contract with incentives that aim the agent to maximize her profit.

Principal's main information about the agent's capacity is deduced from her revenue stream. The revenue consequences of agent's action are referred to as the service performance characteristics, and quantified service performance metrics are referred to as performance measures. The contracts that use performance measures are called performance based contracts. By offering an agent performance based contract, the principal transfers part of her risk regarding revenue to the agent's revenue risk, thus providing incentives for the agent to choose the action desired by the principal. If the performance measure is positively correlated with principal's revenue, a rate of award for each unit of the performance measure, known as the piece rate b, is specified in the contract. If the performance measure is negatively correlated with principal's revenue, a penalty rate for each unit of the performance measure, denoted by p, is specified in the contract.

Under performance based contracts, the agent maximizes his utility based on the scheme proposed by the principal, and the principal maximizes her profit while anticipating the agent's optimizing decision. This scenario is referred to as the *second-best solution* (Hölmstrom 1979). Given a compensation scheme, if the agent's utility is globally concave, the second-best solution can be derived using first order condition of the agent's utility, referred to as the *first-order approach*. If the agent's utility is not globally concave, the first-order approach is generally invalid and alternative approaches have to be used such as converting the agent's utility optimization problem into a convex programming problem (Grossman and Hart 1983).

In our case short unit's downtimes (relative to uptimes) imply a higher revenue for the principal, thus the downtimes and their frequency infer the agent's service performance. The service capacity can only be inferred to by the nature of downtimes, which are unobservable before signing the contract. Therefore moral hazard is of concern with performance based contracts. The performance measure adopted here is based on the unit's downtimes. The downtimes are negatively correlated with principal's revenue, and the agent is charged a penalty p \$ for each unit (seconds, minutes, hours or days) of the performance measure.

In Kim et al. (2010) the profit function of the principal and the utility function of the agent are based on three assumptions. First, the unit is mission-critical and the principal owns one unit. Second, the unit is highly reliable such that the service times are relatively short as compared to the uptimes. Third, the service times are independently and identically distributed, and the distribution has no upper bound on the realization of the service times. This model has two pitfalls: (i) Kim et al. (2010) assume the failures as a Poisson arrival process independent of the service times. It allows for a new failure to occur while the unit is still in a failed state, contradicting that no new failure can occur when in a failed state. (ii) The profit/utility functions describe the total profit/utility during a single contract period assumed finite and normalized to 1. Although the contract period is finite, it contradicts their assumption about the service time distribution with no upper bound on duration of the service time.

Table 2.1 The variables of the model

Variable	Description	Type
η	Agent's risk attitude	Exogenous
r	Unit's revenue rate	Exogenous
λ	Unit's failure rate	Exogenous
c	Marginal rate of capacity cost	Exogenous
w	Agent's compensation rate	Determined by the principal
p	Agent's penalty rate	Determined by the principal
μ	Service capacity	Determined by the agent

To repeat, the failure rate of the equipment unit is a constant λ, the repair time is exponential with a constant repair rate μ (the service capacity is the repair rate), yielding a less general model than Kim et al. (2010). Furthermore, we do not restrict the contract to a period of time, rather, the contract can be dynamic and can be offered and accepted/rejected continuously in time.

The unit's failure rate $\lambda > 0$, the principal's expected revenue rate $r > 0$, and the marginal capacity cost $c > 0$, are exogenous variables. The payment rate w and the penalty rate p are determined by the principal, whereas the service capacity $\mu \geq 0$ is determined by the agent. We denote an exogenous scalar parameter η as preference and intensity indicator for agent's risk attitude: $\eta = 0$ for risk-neutral, $\eta > 0$ for risk-averse, and $\eta < 0$ for risk-seeking.

The seven variables that appear in our model are listed in Table 2.1.

Two performance measures are considered in Kim et al. (2010). The first one is *cumulative downtime* – the sum of downtimes during a finite contract period. The second one is the *average downtime*, which uses the sample average of downtimes during a finite contract period as the performance measure. The two measures provide different incentives for the agent's capacity decisions. In essence, the agent's optimal service capacity behaves non-monotonically with the failure rate when using average downtime, while it is monotonically increasing when using cumulative downtime. This is because average downtime reflects the risk differently compared to cumulative downtime. When the failure rate is higher, the expected number of failures is higher during the finite contract period. For a higher number of failures and the same service capacity, average downtime dilutes the agent's risk by a factor proportional to the square of the number of failures as compared to cumulative downtime, thus provides an incentive for the agent to choose a lower service capacity, leading to reduced service performance. We adopt the steady state probability of the failed state as the sole performance measure, which is the equivalence of cumulative downtime in our undetermined time horizon setting.

The literature on principal-agent setting is extensive in economics since the topic is fundamental to the economic analysis of firms' interdependence via contractual agreements that impact their output. We do not survey here the principal-agent literature. This has been done very well by numerous authors. A partial list includes Ross (1973), Hölmstrom (1979), Stiglitz (1974, 1979), Myerson (1983), Hölmstrom and Milgrom (1987), Fudenberg and Tirole (1990), Maskin and Tirole (1990, 1992), and Bolton and Dewatripont (2005). For analytic and numerical solutions to principal-agent problems see Grossman and Hart (1983) and Guesnerie and Laffont (1984).

2.1 Contractual Relationship Between a Principal and an Agent

When an agent contracts a single principal, the agent is always available when the unit fails, therefore the unit's downtimes are the same as the service times. To mitigate the pitfalls in Kim et al. (2010) we recast this system a Markov process. The state of the Markov process is defined as the state of the principal's unit: in state 0 when the unit is operational, and in state 1 when the unit is not operational. We assume that the uptimes of the unit are independently and identically distributed following an exponential distribution that is governed by the unit's failure rate, and the service times of the unit are independently and identically distributed, following an exponential distribution governed by the agent's service capacity. For a risk-neutral agent we propose an objective function that describes his expected utility rate for each unit of time in an infinite time contract assuming the Markov process is in steady state. Similarly we propose an objective function that describes a risk-neutral principal's expected profit rate. Both the principal's and the agent's objective functions depend on the compensation rate $w > 0$ paid by the principal to the agent and the penalty rate $p > 0$ charged by the principal for each unit of downtime. Furthermore, the principal's expected profit rate also depends on the revenue rate $r > 0$, and the agent's expected utility rate also depends on the marginal cost $c > 0$ of the service capacity for each unit of time. In our principal-agent contractual relationship, the principal controls w and p, and the agent controls μ, therefore we call vector $((w, p), \mu)$ a *strategy*. The c is exogenously determined by the market and in this paper it is normalized as a monetary unit $\Rightarrow c \equiv 1$. Observation 3.1 (below) points out that a contract with compensation rate w paid only for each unit of uptime and penalty rate charged for each unit of downtime is equivalent to our setting of principal-agent contract.

Notation: Denote the principal's expected profit rate by $\Pi_P(w, p; \mu)$ and the agent's expected utility rate by $u_A(\mu; w, p)$, omitting the exogenous parameters.

When the agent does not accept the contract offer he commits no service capacity and receives no compensation. $\underline{u_A}(\mu = 0) = 0$ is referred to as *the agent's reservation utility rate*. An agent accepts the contract only if his expected utility rate is greater than or equal to his reservation utility rate, referred to as the individual rationality (IR) constraints. When the principal does not contract an agent for the repair service, then since an equipment failure will occur after some finite time with probability 1, therefore in the long run the principal's expected profit rate equals zero, which is referred to as *the principal's reservation profit rate* ($\underline{\Pi_P} = 0$).

Individual rationality principal dictates that the principal offers a contract only if her expected profit rate is strictly greater than her reservation profit rate.

When a principal-agent contract exists, the agent's average utility over a finite period of time converges to his expected utility rate as the period approaches infinity. However it is still probable that the agent receives negative revenue stream over some finite period of time, such that his cumulative revenue (utility) drops below a certain threshold and triggers bankruptcy preference claim against the agent. In our paper, we presume that the likelihood of such bankruptcy condition to occur is negligible.

The above principal-agent problem is characterized by expression of the principal's and agent's expected profit/utility rates and the values of the exogenous parameters. Denote a principal-agent problem by $\mathfrak{P}(\Pi_P, u_A, \eta, \lambda, r)$ or for short $\mathfrak{P}$.

Definition 2.1 (*Strategy Set*). *The **strategy set** of a principal-agent problem $\mathfrak{P}$ is defined as a vector $\mathfrak{S}(\mathfrak{P}) \equiv \{((w,p),\mu)\,|w > 0, p > 0, \mu \geq 0\}$.*

Definition 2.2 (*Weak Domination*). *Consider two strategies $((w,p),\mu)$, $((w',p'),\mu') \in \mathfrak{S}(\mathfrak{P})$. $((w,p),\mu)$ is said to **weakly dominates** $((w',p'),\mu')$, denoted by $((w,p),\mu) \succeq ((w',p'),\mu')$, if the two strategies result in $\Pi_P(w,p;\mu) \geq \Pi_P(w',p';\mu')$ and $u_A(\mu;w,p) \geq u_A(\mu';w',p')$ with at least one strict inequality.*

Definition 2.3 (*Set of Admissible Solutions*). *The **set of admissible solutions** (also known as the **set of Pareto optimal solutions**) for the principal-agent problem $\mathfrak{P}$ is the set $\mathfrak{s}(\mathfrak{P})$ of all strategies $((w,p),\mu) \in \mathfrak{S}(\mathfrak{P})$ for which:*

(a) $\nexists\ ((w',p'),\mu') \in \mathfrak{S}(\mathfrak{P})$ *such that* $((w',p'),\mu') \succeq ((w,p),\mu)$ *– there is no other strategy that weakly dominates* $((w,p),\mu)$.
(b) $\Pi_P(w,p;\mu) > \underline{\Pi_P}$ *and* $u_A(\mu;w,p) \geq \underline{u_A}$.

Pareto optimality implies that the principal cannot increase her expected profit rate without lowering the agent's expected utility rate and vice versa (Luenberger 1995), and it has been proven that generally both the principal and the agent achieve Pareto optimality as a subset of the second-best solutions (Ross 1973). Since the agent's IR is always binding, condition (a) in Definition 2.3 guarantees that all admissible solutions are Pareto optimal. We require that all the solutions proposed in this paper be *Admissible Solutions*.

This paper is organized as follows. In Chap. 3, we present the basic model with a risk-neutral principal and a risk-neutral agent, and we describe the exogenous conditions that guarantee the existence of a contract and the optimal contract terms. In Chap. 4 we analyze risk-averse agent. Chapter 5 is dedicated to the analysis of a risk-seeking agent. In Chap. 6 we summarize our findings and conclusions. Notation is introduced as needed.

Chapter 3
Risk-Neutral Agent

When a risk-neutral agent accepts a contract offer (w, p), his expected utility rate is composed of the expected value of the compensation rate from the principal and a deterministic cost rate of the service capacity which can be expressed as $w - pP(1) - \mu$, where $P(1)$ denotes the steady state probability of the unit being in the failed state. Similarly denote the steady state probability of the unit being operational by $P(0) = 1 - P(1)$.

Notation: $(x)_+ = x$ when $x \geq 0$ and $(x)_+ = 0$ when $x < 0$.

A risk-neutral agent's expected utility rate is:

$$u_A(\mu; w, p) = (w - pP(1) - \mu)_+ \quad \text{for } w > 0, p > 0, \mu \geq 0 \tag{3.1}$$

$P(0)$ and $P(1)$ (functions of λ and μ), represent the proportion of time in the steady state the Markov process is in state 0 and state 1 respectively (Ross 2006). They satisfy the balance equations of the Markov process and sum up to 1, thus $P(0) = \mu/(\lambda + \mu)$, $P(1) = \lambda/(\lambda + \mu)$:

$$u_A(\mu; w, p) = \left(w - \frac{p\lambda}{\lambda + \mu} - \mu\right)_+ \quad \text{for } w > 0, p > 0, \mu \geq 0 \tag{3.2}$$

Since the principal determines w and p she can always entice the agent to accept the contract.

For $r > 0$ (determined exogenously by the market), the principal's expected profit rate is composed of the expected revenue rate generated by her unit, the expected penalty rate collected from the agent and the compensation rate paid to the agent:

S. Zeng, M. Dror, *Formulating Principal-Agent Service Contracts for a Revenue Generating Unit*, SpringerBriefs in Operations Management,
DOI 10.1007/978-3-319-18672-6_3

$$\Pi_P(w, p; \mu) = rP(0) - w + pP(1) = \frac{r\mu}{\lambda + \mu} - w + \frac{p\lambda}{\lambda + \mu}$$
$$\text{for } w > 0, p > 0, \mu \geq 0 \tag{3.3}$$

Observation 3.1. *We note that under another type of contract, where the principal compensates the agent only for each unit of uptime (instead of each unit of time), the agent's expected utility rate is equivalent to* (3.2)*, and the principal's expected profit rate is equivalent to* (3.3)*: Under the new type of contract, denote the compensation rate by* $\tilde{w}$ *and the penalty rate by* $\tilde{p}$*, therefore the agent's expected utility rate becomes:*

$$u_A(\mu; \tilde{w}, \tilde{p}) = (\tilde{w}P(0) - \tilde{p}P(1) - \mu)_+ = \left(\frac{\tilde{w}\mu}{\lambda + \mu} - \frac{\tilde{p}\lambda}{\lambda + \mu} - \mu\right)_+$$
$$\text{for } \tilde{w} > 0, \tilde{p} > 0, \mu \geq 0 \tag{3.4}$$

and the principal's expected profit rate becomes:

$$\Pi_P(\tilde{w}, \tilde{p}; \mu) = rP(0) - \tilde{w}P(0) + \tilde{p}P(1) = \frac{r\mu}{\lambda + \mu} - \frac{\tilde{w}\mu}{\lambda + \mu} + \frac{\tilde{p}\lambda}{\lambda + \mu}$$
$$\text{for } \tilde{w} > 0, \tilde{p} > 0, \mu \geq 0 \tag{3.5}$$

Replacing $\tilde{w}$ *by* w *and* $\tilde{p}$ *by* $(p - w)$ *in* (3.4) *and* (3.5) *we obtain* (3.2) *and* (3.3) *respectively.*

Note that a performance based contract can even take the form such that a compensation rate is specified for each unit of uptime (instead of each unit of time) and no penalty rate is charged whatsoever. That is, the principal controls only one variable (the compensation rate) instead of two (the compensation rate and the penalty rate). However this form of performance based contract is not discussed in this work.

Returning to the agent as in (3.2) we define the part inside the brackets by

$$u(\mu) \equiv w - \frac{p\lambda}{\lambda + \mu} - \mu \tag{3.6}$$

i.e., for $\mu \geq 0$, $u(\mu)$ is continuous and differentiable everywhere:

$$\frac{du(\mu)}{d\mu} = \frac{p\lambda}{(\lambda + \mu)^2} - 1 \text{ and } \frac{d^2u(\mu)}{d\mu^2} = -\frac{2p\lambda}{(\lambda + \mu)^3} < 0$$

$$u(0) = w - p, \quad \left.\frac{du(\mu)}{d\mu}\right|_{\mu=0} = \frac{p}{\lambda} - 1 \text{ and } \lim_{\mu \to +\infty} \frac{du(\mu)}{d\mu} = -1$$

3.1 Optimal Strategies for Risk-Neutral Agent

Note that $u(\mu)$ in (3.6) increases and $\Pi_P(w, p; \mu)$ in (3.3) decreases in w, therefore for any value of penalty rate p, the principal can raise her expected profit rate by adjusting the rate w low enough while ensuring the agent's participation by setting the agent's expected utility rate equal to his reservation utility rate. Although the principal cannot contract directly on the agent's capacity, she presumes the agent will optimize his expected utility rate. That is, for any compensation rate w and penalty rate p proposed by the principal, the agent computes the value of μ that maximizes his expected utility rate and decides whether to accept the contract or not by solving the following optimization problem:

$$\max_{\mu \geq 0} u(\mu) = \max_{\mu \geq 0} \left\{ w - \frac{p\lambda}{\lambda + \mu} - \mu \right\} \tag{3.7}$$

with agent's optimal service capacity denoted by $\mu^*(w, p) = \text{argmax}_{\mu \geq 0} u(\mu)$.

We describe the agent's optimal response to any possible contract offer $(w, p) \in \mathbb{R}^2_+$ in Proposition 3.3, but we start with a simple technical lemma – one of many.

Lemma 3.2. *If $p > \lambda > 0$, then $p > 2\sqrt{p\lambda} - \lambda > 0$.*

Proof. If $p > \lambda > 0$, then $2\sqrt{p\lambda} - \lambda > 2\lambda - \lambda = \lambda > 0$ and $p - 2\sqrt{p\lambda} + \lambda = \left(\sqrt{p} - \sqrt{\lambda}\right)^2 > 0$, where the latter inequality indicates $p > 2\sqrt{p\lambda} - \lambda$. □

Proposition 3.3. *Consider a risk-neutral agent with $u_A(\mu; w, p)$ given in* (3.2).

(a) *Given $p \in (0, \lambda]$, then the agent accepts the contract only when $w \geq p$ and does not commit any service capacity ($\mu^*(w, p) = 0$) resulting in expected utility rate $u_A(\mu^*(w, p); w, p) = w - p \geq 0$.*
(b) *Given $p > \lambda$, then the agent accepts the contract only when $w \geq 2\sqrt{p\lambda} - \lambda$ and installs service capacity $\mu^*(w, p) = \sqrt{p\lambda} - \lambda > 0$ resulting in expected utility rate $u_A(\mu^*(w, p); w, p) = w - 2\sqrt{p\lambda} + \lambda \geq 0$.*

Proof. Figure 3.1 illustrates the form of $u(\mu)$ when the value of p falls in different ranges. The structure of the proof for Proposition 3.3 is depicted in Fig. 3.2.

Case $p \in (0, \lambda]$: $u(\mu)$ is decreasing for $\mu \geq 0$, therefore the optimal service capacity is set at $\mu^*(w, p) = 0$ and $u(\mu^*(w, p)) = w - p$.

Subcase $w \in (0, p)$: $u(\mu^*(w, p)) < 0$, therefore the agent rejects the contract.
Subcase $w \geq p$: $u(\mu^*(w, p)) \geq 0$, thus the agent would accept the contract if offered.

Case $p > \lambda$: The service capacity that maximizes $u(\mu)$ is positive as seen from the first order condition $du(\mu)/d\mu|_{\mu=\mu^*(w,p)} = 0 \Rightarrow \mu^*(w, p) = \sqrt{p\lambda} - \lambda > 0$ and $u(\mu^*(w, p)) = w - 2\sqrt{p\lambda} + \lambda$. According to Lemma 3.2 we have to resolve the following subcases:

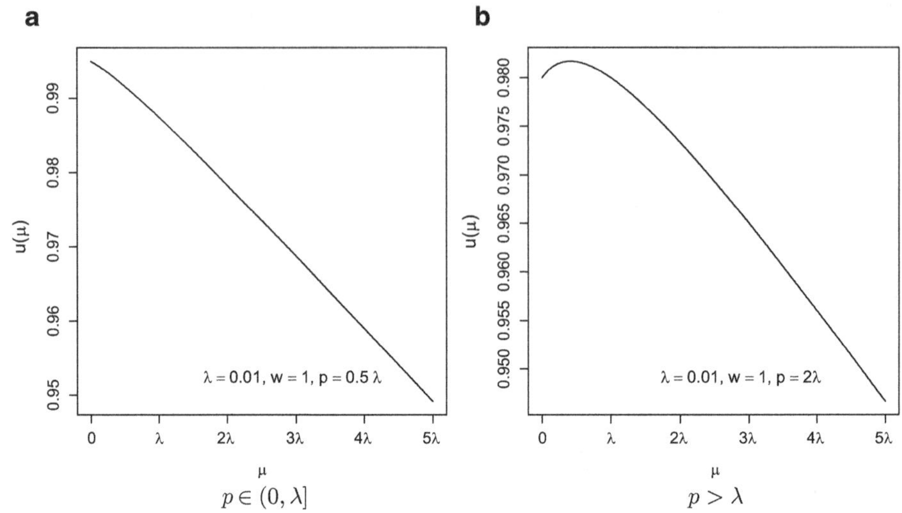

Fig. 3.1 Illustration of the forms of $u(\mu)$

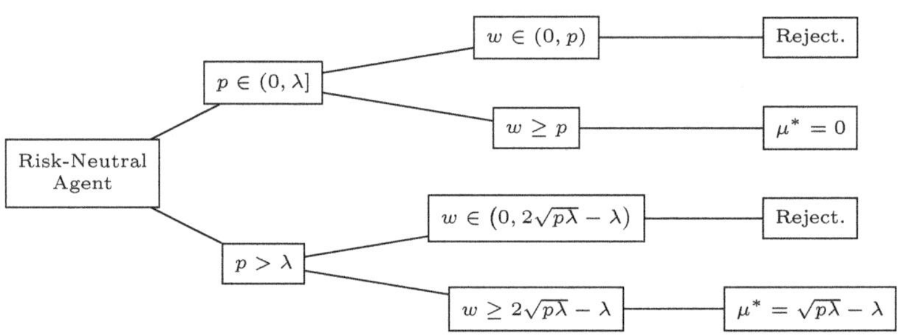

Fig. 3.2 Structure of the proof for Proposition 3.3

Subcase $w \in \left(0, 2\sqrt{p\lambda} - \lambda\right)$: $u(\mu^*(w,p)) < 0$, therefore the agent rejects the contract.

Subcase $w \geq 2\sqrt{p\lambda} - \lambda$: $u(\mu^*(w,p)) \geq 0$, therefore the agent would accept the contract if offered.

□

In summary, given exogenous market conditions such that there exists a contract benefiting both the agent and principal (see Theorem 3.4 later), only one formula is necessary for the agent to determine his service capacity: $\mu^*(w,p) = \sqrt{p\lambda} - \lambda > 0$.

The conditions when the agent accepts the contract are depicted by the shaded areas in Fig. 3.3. The two shaded areas of different grey scales represent conditions $\{(w,p) : p \in (0, \lambda], w \geq p\}$ and $\left\{(w,p) : p > \lambda, w \geq 2\sqrt{p\lambda} - \lambda\right\}$ under which the agent accepts the contract but responds differently. The lower bound function of the

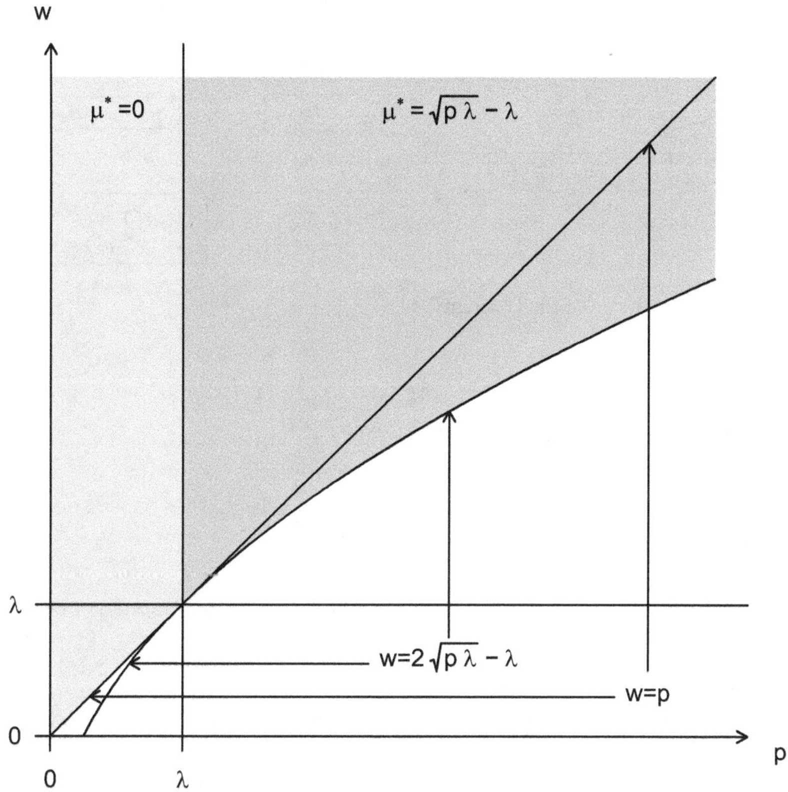

Fig. 3.3 Conditions when a risk-neutral agent accepts the contract

shaded areas (denoted by $w_0(p)$) represents the contract offers that result in agent zero expected utility rate. $w_0(p)$ is defined as follows:

$$w_0(p) = \begin{cases} p \text{ for } p \in (0, \lambda] \\ 2\sqrt{p\lambda} - \lambda \text{ for } p > \lambda \end{cases}$$

Note that since $\lim_{p\to\lambda^-} w_0(p) = \lim_{p\to\lambda^+} w_0(p) = \lambda$, $\lim_{p\to\lambda^-} dw_0(p)/dp = \lim_{p\to\lambda^+} dw_0(p)/dp = 1$, $w_0(p)$ is continuous and differentiable everywhere for $p \in \mathbb{R}_+$.

Anticipating (calculating) the agent's optimal response $\mu^*(w, p)$ the principal chooses w and p that maximize her expected profit rate by solving the optimization problem:

$$\max_{w>0,p>0} \Pi_P(w, p; \mu^*(w, p)) = \max_{w>0,p>0} \left\{ \frac{r\mu^*(w, p)}{\lambda + \mu^*(w, p)} - w + \frac{p\lambda}{\lambda + \mu^*(w, p)} \right\} \tag{3.8}$$

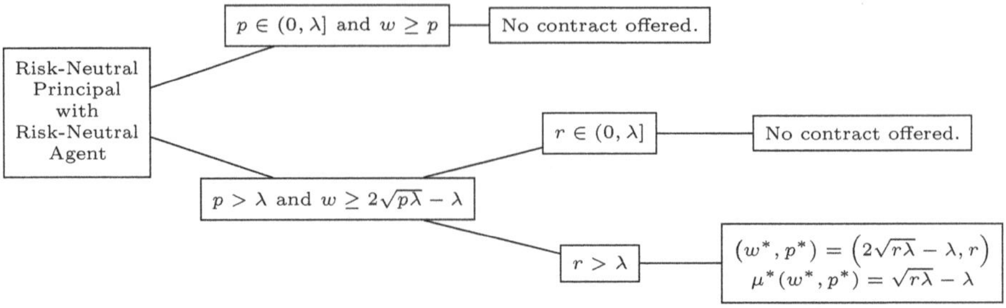

Fig. 3.4 Structure of the proof for Theorem 3.4

with the optimal rates $(w^*, p^*) = \text{argmax}_{w>0, p>0} \Pi_P(w, p; \mu^*(w, p))$. We only consider pairs $(w, p) \in \mathbb{R}_+^2$ such that $u_A(\mu; w, p) \geq 0$.

Define: $$\mathfrak{D}_{RN} \equiv \{(w, p) : p \in (0, \lambda], w \geq p\} \cup \left\{(w, p) : p > \lambda, w \geq 2\sqrt{p\lambda} - \lambda\right\} \tag{3.9}$$

Theorem 3.4. *Given a risk-neutral agent as in* (3.2) *and a principal as in* (3.3) *and suppose that* $(w, p) \in \mathfrak{D}_{RN}$.

(a) *If* $r \in (0, \lambda]$, *then the principal does not propose a contract.*
(b) *If* $r > \lambda$, *then the principal's offer and the agent's capacity are respectively*

$$\left(w^*, p^*\right) = \left(2\sqrt{r\lambda} - \lambda, r\right) \text{ and } \mu^*(w^*, p^*) = \sqrt{r\lambda} - \lambda \tag{3.10}$$

resulting in principal's expected profit rate $\Pi_P(w^*, p^*; \mu^*(w^*, p^*)) = r - 2\sqrt{r\lambda} + \lambda$.

Proof. The structure of the proof for Theorem 3.4 is depicted in Fig. 3.4.

Case $p \in (0, \lambda]$ and $w \geq p$: According to Proposition 3.3 part (a), the agent would accept the contract without installing any service capacity. Since $\partial \Pi_P / \partial w = -1 < 0$, the principal chooses $w^* = p$ and $\Pi_P(w^*, p; \mu^*(w^*, p)) = -p + p = 0$. Left with zero expected profit rate, the principal does not propose a contract.

Case $p > \lambda$ and $w \geq 2\sqrt{p\lambda} - \lambda$: According to Proposition 3.3 part (b), the agent accepts the contract and installs capacity $\sqrt{p\lambda} - \lambda$. Since $\partial \Pi_P / \partial w = -1 < 0$, therefore $w^* = 2\sqrt{p\lambda} - \lambda$ and the principal's optimization problem becomes $\max_{p>\lambda} \Pi_P(w^*, p; \mu^*(w^*, p))$ where:

$$\Pi_P(w^*, p; \mu^*(w^*, p)) = r + \lambda - \sqrt{\lambda}\left(\sqrt{p} + \frac{r}{\sqrt{p}}\right) \tag{3.11}$$

Define $x \equiv \sqrt{p}, a \equiv \sqrt{\lambda}$. The principal's expected profit rate, denoted by $f(x)$, can be restated as $f(x) = r + a^2 - a\,(x + r/x)$ for $x > 0$ and $a > 0$. Maximizing $f(x)$ with respect to $x > 0$ is equivalent to maximizing $\Pi_P(w^*, p; \mu^*(w^*, p))$ with respect to $p > 0$ in the sense that

$$\underset{p>0}{\text{argmax}}\ \Pi_P(w^*, p; \mu^*(w^*, p)) = \left(\underset{x>0}{\text{argmax}} f(x)\right)^2$$

Denote $p^\star \equiv \text{argmax}_{p>0}\, \Pi_P(w^*, p; \mu^*(w^*, p))$. Since $d^2 f(x)/dx^2 = -2ar/x^3 < 0$, therefore $f(x)$ is concave with respect to $x > 0$ and from the first order condition $df(x)/dx|_{x=x^*} = ar/\,(x^*)^2 - a = 0 \Rightarrow x^* = \sqrt{r}$. Therefore $p^\star = (x^*)^2 = r$. However $p^\star = r$ is not necessarily the optimal solution because the principal maximizes p for $p > \lambda$. Thus $p^* = \max\{r, \lambda\}$.

Subcase $r \in (0, \lambda]$: $p^* = \lambda$; the principal does not propose a contract since her expected profit rate is zero.

Subcase $r > \lambda$: $p^* = r$; the principal receives $\Pi_P(w^*, p^*; \mu^*(w^*, p^*)) = r - 2\sqrt{r\lambda} + \lambda = \left(\sqrt{r} - \sqrt{\lambda}\right)^2 > 0$ and proposes a contract $(w^*, p^*) = (2\sqrt{r\lambda} - \lambda, r)$ that induces the agent to install service capacity $\mu^*(w^*, p^*) = \sqrt{r\lambda} - \lambda$.

In summary, if $r \in (0, \lambda]$, then the principal does not propose a contract (Theorem 3.4 (a)). If $r > \lambda$, then the principal offers $(w^*, p^*) = \left(2\sqrt{r\lambda} - \lambda, r\right)$ and the agent installs capacity $\mu^*(w^*, p^*) = \sqrt{r\lambda} - \lambda$ (Theorem 3.4 (b)), which is an admissible solution according to Definition 2.3. □

Note that in an optimal contract configuration the agent compensates fully the principal for lost revenue during the unit's fail duration.

3.1.1 Sensitivity Analysis of the Optimal Strategy

The principal-agent rationality assumption are odds with the agent accepting a contract offer and responding with $\mu^* = 0$. Therefore the only viable case is when the agent accepts the contract and installs $\mu^*(w, p) = \sqrt{p\lambda} - \lambda$. In this case the rate w is bounded below by $2\sqrt{p\lambda} - \lambda = pP(1) + \mu^*(w, p)$, with $pP(1)$ representing the expected penalty rate charged by the principal when the optimal capacity is installed. It implies that the agent should at least be reimbursed for the expected penalty rate and the cost of the optimal service capacity in exchange for his repair service.

The optimal service capacity itself depends only on p and λ. Note that $\partial\mu^*/\partial p = \sqrt{\lambda/4p} > 0$ and $\partial\mu^*/\partial\lambda = \sqrt{p/4\lambda} - 1$. It indicates that given a λ the agent will increase the μ when the p increases. However, given a p the change in μ^* with respect to the failure rate is not monotonic. The $\sqrt{p\lambda} - \lambda$, as a function of λ, increases when $\lambda \in (0, p/4)$ and decreases when $\lambda \in (p/4, p)$. If the principal's unit is reliable ($\lambda \in (0, p/4)$), then the agent increases the μ when λ increases. If the principal's unit is less reliable ($\lambda \in (p/4, p)$), then the savings from reducing the μ are greater than the increase in p, therefore the agent will reduce μ^* when the λ increases.

The agent's optimal expected utility rate when installing capacity $\mu^*(w,p) = \sqrt{p\lambda} - \lambda$ is $u_A^* \equiv u_A(\mu^*(w,p); w,p) = w - 2\sqrt{p\lambda} + \lambda$, and it depends on w, p and λ. Note that $\partial u_A^*/\partial w = -1 < 0$, $\partial u_A^*/\partial p = -\sqrt{\lambda/p} < 0$, indicating that the agent's optimal expected utility rate decreases with the compensation rate and the penalty rate. Note that $\partial u_A^*/\partial \lambda = -\sqrt{p/\lambda} + 1$, and from Proposition 3.3 $p > \lambda \Rightarrow -\sqrt{p/\lambda} + 1 < 0$, therefore the agent's optimal expected utility rate also decreases with the failure rate.

According to Theorem 3.4, a principal offers a contract to a risk-neutral agent only if $r > \lambda$ and her offer is $(w^*, p^*) = \left(2\sqrt{r\lambda} - \lambda, r\right)$ with expected profit rate $\Pi_P^* \equiv \Pi_P(w^*, p^*; \mu^*(w^*, p^*)) = r - 2\sqrt{r\lambda} + \lambda = \left(\sqrt{r} - \sqrt{\lambda}\right)^2$. The compensation rate and the expected profit rate depend on r and λ, and the penalty rate equals r. Note that $\partial w^*/\partial r = \sqrt{\lambda/r} > 0$ and $\partial w^*/\partial \lambda = \sqrt{r/\lambda} - 1 > 0$ implying that given the λ, the principal will increase w when the revenue rate increases, and given the revenue rate, the principal will increase w when λ increases. Note that $\partial \Pi_P^*/\partial r = \left(\sqrt{r} - \sqrt{\lambda}\right)/\sqrt{r} > 0$ and $\partial \Pi_P^*/\partial \lambda = -\left(\sqrt{r} - \sqrt{\lambda}\right)/\sqrt{\lambda} < 0$. These results imply that given λ, principal's expected profit rate will increase when the revenue rate increases, and given the revenue rate, principal's expected profit rate will decrease when her equipment unit becomes less reliable.

3.1.2 The Second-Best Solution

According to Theorem 3.4, $((w^*, p^*) = \left(2\sqrt{r\lambda} - \lambda, r\right), \mu^*(w^*, p^*) = \sqrt{r\lambda} - \lambda)$ is the second-best solution. When the principal can contract directly on μ there is no moral hazard. Therefore in first-best setting, the agent's expected utility rate, denoted by $u_A^{FB}(w, \mu)$, is simply $u_A^{FB}(w, \mu) = (w - \mu)_+$ for $w > 0$ and $\mu > 0$. Since the principal determines w and μ, her optimization problem is:

$$\max_{w>0,\mu>0} \Pi_P^{FB}(w, \mu) = \max_{w>0,\mu>0} \{rP(0) - w\} = \max_{w>0,\mu>0} \left\{\frac{r\mu}{\lambda + \mu} - w\right\} \tag{3.12}$$

Denote w^{FB} and μ^{FB} the corresponding solution. Since $\partial \Pi_P^{FB}/\partial w = -1 < 0$, therefore the principal chooses $w^{FB} = \mu$ to ensure the agent's participation and her optimization problem becomes:

$$\max_{\mu>0} \Pi_P^{FB}(\mu) = \max_{\mu>0} \left\{\frac{r\mu}{\lambda + \mu} - \mu\right\} \tag{3.13}$$

Since $d^2\Pi_P^{FB}(\mu)/d\mu^2 = -2r\lambda/(\lambda+\mu)^3 < 0$, the principal's expected profit rate is concave with respect to $\mu > 0$ and μ^{FB} can be derived from the first order condition $d\Pi_P^{FB}(\mu)/d\mu\big|_{\mu=\mu^{FB}} = r\lambda/\left(\lambda+\mu^{FB}\right)^2 - 1 = 0 \Rightarrow \mu^{FB} = \sqrt{r\lambda} - \lambda$. However $\mu^{FB} = \sqrt{r\lambda} - \lambda$ may not necessarily be the optimal solution because the principal requires $\mu > 0$. Note that $\mu^{FB} = \sqrt{\lambda}\left(\sqrt{r} - \sqrt{\lambda}\right) > 0$ only if $r > \lambda$. Therefore the first-best solution is:

$$w^{FB} = \mu^{FB} = \sqrt{r\lambda} - \lambda \text{ for } r > \lambda \tag{3.14}$$

By comparing the second-best solution (3.10) to the first-best solution (3.14), we conclude:

1. The principal offers a contract only when $r > \lambda$ indicating that the existence of a beneficial contract for risk-neutral agent is determined exogenously by the market (the revenue rate r) and the nature of the equipment (the failure rate λ), which is consistent with Proposition 2 in Harris and Raviv (1978).
2. The proposed w in the second-best solution is higher than that in the first-best solution ($w^* = 2\sqrt{r\lambda} - \lambda > \sqrt{r\lambda} - \lambda = w^{FB}$), because the principal has to compensate for the p when the agent's μ is not observable. Nevertheless, the second-best contract is efficient (as the first-best contract) because of point 3 below.
3. The optimal capacity in the first-best solution and the second-best solution are the same ($\mu^{FB} = \mu^*(w^*, p^*) = \sqrt{r\lambda} - \lambda$), indicating that the principal can induce a risk-neutral agent to install the desired capacity without contracting on it directly. Furthermore, the principal receives the same expected profit rate no matter if the agent's action is observable (thus contractible) or not. This is consistent with Proposition 3 part (i) in Harris and Raviv (1978).
4. Finally when the agent is risk-neutral, the principal is guaranteed getting the revenue rate r at all times regardless of the state of the equipment unit (because $p^* = r$). This comes at the cost of the contract ($w^* = 2\sqrt{r\lambda} - \lambda$). In other words, the principal's profit rate appears as if it is deterministic. However this is not true for a risk-averse agent, as seen in Chap. 4.

3.1.3 Our Principal-Agent Game

To clarify the interplay of decisions by the principal and the agent, we cast the principal-agent problem in an extensive form game depicted in Fig. 3.5 below, where "P" represents the principal and "A" the agent.

There are four possible strategies the principal can choose from:

O_1: Offer a contract with $p \in (0, \lambda]$ and $w \in (0, p)$.
O_2: Offer a contract with $p \in (0, \lambda]$ and $w \geq p$.

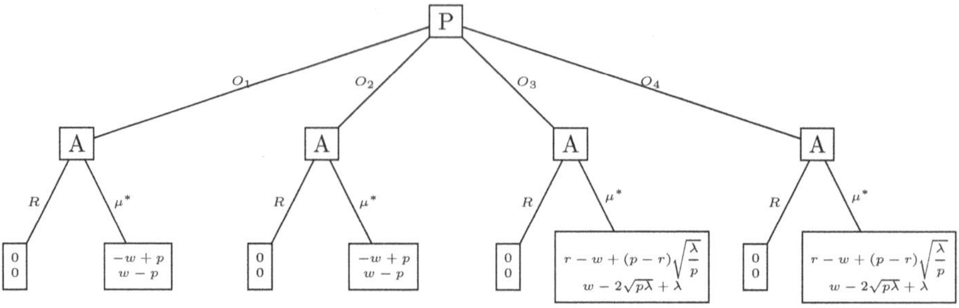

Fig. 3.5 Structure of the principal-agent extensive form game

O_3: Offer a contract with $p > \lambda$ and $w \in \left(0, 2\sqrt{p\lambda} - \lambda\right)$.

O_4: Offer a contract with $p > \lambda$ and $w \geq 2\sqrt{p\lambda} - \lambda$.

For any contract offer by the principal, there are two strategies for the agent to choose from: "*R*" when rejecting the contract, and μ^* for accepting the contract and installing the service capacity that maximizes the agent's expected profit rate. If the principal offers O_1 or O_2 and the agent accepts the contract, then $\mu^* = 0$. If the principal offers O_3 and O_4 and the agent accepts the contract, then $\mu^* = \sqrt{p\lambda} - \lambda$.

The principal's expected profit rate and the agent's expected utility rate are presented in the leaves of the tree in Fig. 3.5. The element above and below are the principal's and the agent's values respectively.

The agent would accept the contract only if his maximized expected utility rate is no less than his reservation utility rate $\underline{u_A} = 0$, therefore the agent accepts the contract when the principal offers O_2 and O_4, and rejects the contract when the principal offers O_1 and O_3. The principal always prefers the agent to accept the contract and install a positive service capacity. Therefore the principal would choose O_4 to all other options. Thus there is only one (subgame perfect) Nash equilibrium: the principal offers a contract with $p > \lambda$ and $w \geq 2\sqrt{p\lambda} - \lambda$ and the agent accepts the contract and installs $\mu^* = \sqrt{p\lambda} - \lambda > 0$.

Chapter 4
Risk-Averse Agent

What if the agent is risk-averse. Fluctuations of the agent's revenue stream occur because the principal's equipment unit can be either in state 0 ('operational') or in state 1 ('down'). In the operational state the penalty rate is 0, whereas in the down state the penalty rate is p. In other words, the penalty rate at any point of time can be modeled as pB where B is a Bernoulli random variable of value 0 with probability $P(0) = \mu/(\lambda+\mu)$ and value 1 with probability $P(1) = \lambda/(\lambda+\mu)$. The dispersion of B decreases as $P(1)$ moves away from $1/2$ in either direction. Denote momentarily $a \equiv P(1)$.

The risk of a random variable is often expressed by the dispersion of the underlying random fluctuation. Standard deviation is commonly used to measure the dispersion of revenue in risk sharing contracts because it is conveniently additive with the revenue stream (Stiglitz 1974; Fukunaga and Huffman 2009; Lewis and Bajari 2014). The standard deviation of pB as a function of a, is denoted by

$$s(a) \equiv \sigma_{pB} = p\sqrt{a(1-a)} \text{ for } a \in [0, 1]$$

We have modified the above risk measure somewhat. Since $s(a)$ strictly decreases as a moves away from $1/2$ in either direction so any other dispersion measure of pB that has this property is a monotone increasing function of the standard deviation $s(a)$. We choose to adopt the dispersion measure:

$$r(a) \equiv p\left(\frac{1}{2} - \left|\frac{1}{2} - a\right|\right) \text{ for } a \in [0, 1]$$

The $r(a)$ above is strictly decreasing as a gets away from $1/2$ in either direction and $r(a)$ has the property that for any $a, a' \in [0, 1]$, we have

$$r(a) \leq r(a') \Leftrightarrow s(a) \leq s(a')$$

S. Zeng, M. Dror, *Formulating Principal-Agent Service Contracts for a Revenue Generating Unit*, SpringerBriefs in Operations Management,
DOI 10.1007/978-3-319-18672-6_4

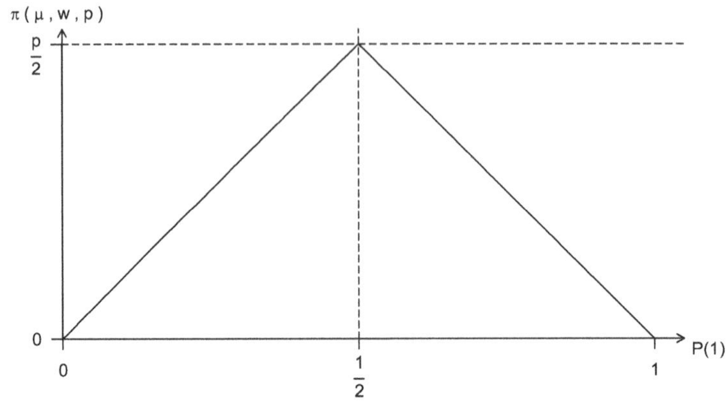

Fig. 4.1 $\pi(\mu, w, p)$ as a function of $P(1)$ when $\eta = 1$

Note that $r(a)$ increases (decreases) if and only if the standard deviation $s(a)$ increases (decreases).

Risk premium of a risk-averse agent is the $ value he is willing to forfeit to avoid uncertainties (fluctuations) in his revenue stream and as a consequence the risk premium is defined as follows:

$$\pi(\mu, w, p) = \eta p\left(\frac{1}{2} - \left|\frac{1}{2} - a\right|\right) = \eta p\left(\frac{1}{2} - \left|\frac{1}{2} - P(1)\right|\right)$$
$$= \eta p\left(\frac{1}{2} - \left|\frac{1}{2} - \frac{\lambda}{\lambda + \mu}\right|\right) \tag{4.1}$$

Figure 4.1 is an example that depicts the shape of $\pi(\mu, w, p)$ as a function of $P(1)$ when $\eta = 1$. $\pi(\mu, w, p)$ reaches its peak when the equipment has equal likelihood of being operational and being failed. In such case the agent can hardly infer anything from the state of the equipment in order to predict his revenue stream and therefore it is considered the most risky. When the likelihood of the equipment being operational is close to 1, the agent can predict his revenue stream more precisely (less risky). Similarly when the likelihood of the equipment being failed is close to 1, the agent can also predict his revenue stream more precisely.

The real parameter η indicates the preference and intensity of the agent's risk attitude. When $\eta > 0$ the agent is risk-averse, when $\eta = 0$ the agent is risk-neutral (and the model reduces to the model of Chap. 3), and $\eta < 0$ indicates that the agent is risk-seeking (see Chap. 5). In the analysis below, the value η plays the role of an exogenous variable.

Modifying (3.2), the risk-averse agent's expected utility rate in this section is:

$$u_A(\mu; w, p) = \left(w - \frac{p\lambda}{\lambda + \mu} - \mu - \eta p \left(\frac{1}{2} - \left| \frac{1}{2} - \frac{\lambda}{\lambda + \mu} \right| \right) \right)_+$$
$$\text{for } w > 0, p > 0, \mu \geq 0 \tag{4.2}$$

Note that $\eta > 0 \Rightarrow \pi(\mu, w, p) \geq 0$, and such a risk premium being subtracted from a risk-neutral agent's expected utility rate (as in (4.2)) implies risk-aversion. The analysis is different for $\eta \in (0, 4/5)$ compared to $\eta \geq 4/5$. Thus, for convenience, when $\eta \in (0, 4/5)$ we describe the agent as *weakly risk-averse*, and when $\eta \geq 4/5$ we describe the agent as *strongly risk-averse*. We assume, say for historical reasons, that both the agent and the principal know not only the type of the risk-averse agent, but also the value of η.

The principal is always risk-neutral and her expression of expected profit rate $\Pi_P(w, p; \mu)$ is the same as (3.3).

Define the part inside the brackets in (4.2) as

$$u(\mu) \equiv w - \frac{p\lambda}{\lambda + \mu} - \mu - \eta p \left(\frac{1}{2} - \left| \frac{1}{2} - \frac{\lambda}{\lambda + \mu} \right| \right)$$
$$= \begin{cases} w - \eta p - \dfrac{(1 - \eta)p\lambda}{\lambda + \mu} - \mu, \; \mu \in [0, \lambda] \\ w - \dfrac{(1 + \eta)p\lambda}{\lambda + \mu} - \mu, \; \mu > \lambda \end{cases} \tag{4.3}$$

The behavior of the utility function $u(\mu)$ for $\mu \geq 0$ is of prime technical interest. Note that $u(\mu)$ is differentiable everywhere on $\mu \geq 0$ except at $\mu = \lambda$. When $\mu \in [0, \lambda)$:

$$\frac{du(\mu)}{d\mu} = \frac{(1 - \eta)p\lambda}{(\lambda + \mu)^2} - 1, \quad \lim_{\mu \to 0+} \frac{du(\mu)}{d\mu} = \frac{1 - \eta}{\lambda} \left(p - \frac{\lambda}{1 - \eta} \right)$$

$$\lim_{\mu \to \lambda-} \frac{du(\mu)}{d\mu} = \frac{1 - \eta}{4\lambda} \left(p - \frac{4\lambda}{1 - \eta} \right) \text{ and } \frac{d^2u(\mu)}{d\mu^2} = -\frac{2(1 - \eta)p\lambda}{(\lambda + \mu)^3}$$

and when $\mu > \lambda$:

$$\frac{du(\mu)}{d\mu} = \frac{(1 + \eta)p\lambda}{(\lambda + \mu)^2} - 1, \quad \lim_{\mu \to \lambda+} \frac{du(\mu)}{d\mu} = \frac{1 + \eta}{4\lambda} \left(p - \frac{4\lambda}{1 + \eta} \right)$$

$$\lim_{\mu \to +\infty} \frac{du(\mu)}{d\mu} = -1 \text{ and } \frac{d^2u(\mu)}{d\mu^2} = -\frac{2(1 + \eta)p\lambda}{(\lambda + \mu)^3} < 0$$

The above derivatives indicate the direction of monotonicity and the concavity/convexity of function $u(\mu)$ over $[0, \lambda)$ and $(\lambda, +\infty)$. Table 4.1 summarizes

Table 4.1 Indicators of the monotonicity and the concavity/convexity of function $u(\mu)$ in (4.3)

Case		Over $[0, \lambda)$ $u_\mu(0^+)$	$u(\mu)$ is	$u_\mu(\lambda^-)$	Over $(\lambda, +\infty)$ $u_\mu(\lambda^+)$	$u(\mu)$ is	$u_\mu(+\infty)$
$\eta \in \left(0, \frac{3}{5}\right]$	$p \in \left(0, \frac{\lambda}{1-\eta}\right]$	≤ 0	Concave	< 0	< 0	Concave	< 0
	$p \in \left(\frac{\lambda}{1-\eta}, \frac{4\lambda}{1+\eta}\right]$[a]	> 0	Concave	< 0	≤ 0	Concave	< 0
	$p \in \left(\frac{4\lambda}{1+\eta}, \frac{4\lambda}{1-\eta}\right]$	> 0	Concave	≤ 0	> 0	Concave	< 0
	$p \in \left(\frac{4\lambda}{1-\eta}, +\infty\right)$	> 0	Concave	> 0	> 0	Concave	< 0
$\eta \in \left(\frac{3}{5}, 1\right)$	$p \in \left(0, \frac{4\lambda}{1+\eta}\right]$	< 0	Concave	< 0	≤ 0	Concave	< 0
	$p \in \left(\frac{4\lambda}{1+\eta}, \frac{\lambda}{1-\eta}\right]$[b]	≤ 0	Concave	< 0	> 0	Concave	< 0
	$p \in \left(\frac{\lambda}{1-\eta}, \frac{4\lambda}{1-\eta}\right]$	> 0	Concave	≤ 0	> 0	Concave	< 0
	$p \in \left(\frac{4\lambda}{1-\eta}, +\infty\right)$	> 0	Concave	> 0	> 0	Concave	< 0
$\eta \in [1, +\infty)$	$p \in \left(0, \frac{4\lambda}{1+\eta}\right]$	< 0	Convex	< 0	≤ 0	Concave	< 0
	$p \in \left(\frac{4\lambda}{1+\eta}, +\infty\right)$	< 0	Convex	< 0	> 0	Concave	< 0

[a]Note that $\eta \in (0, 3/5] \Rightarrow 4\lambda/(1+\eta) \geq \lambda/(1-\eta)$
[b]Note that $\eta \in (3/5, 1) \Rightarrow \lambda/(1-\eta) > 4\lambda/(1+\eta)$

these indicators for various regions of the space $\mathbb{R}_+^2$ of the pairs (η, p). In the table $u_\mu(\cdot) = \lim_{\mu \to (\cdot)} du/d\mu$, and $u_\mu(\cdot^+)$ represents the limit of $u_\mu(\mu)$ as μ approaches $(\cdot)$ from above, and similar for $u_\mu(\cdot^-)$.

4.1 Optimal Strategies with a Weakly Risk-Averse Agent

Similarly to the risk-neutral agent case, agent's expected utility rate increases and principal's expected profit rate decreases in w, therefore for any value of p the principal maximizes her expected profit rate by lowering the compensation rate w yet maintaining the agent's participation by setting the agent's expected utility rate equal to his reservation utility rate. Although the principal cannot contract directly on the agent's service capacity, she anticipates the agent to optimize his expected utility rate when offered a contract. That is, for any w and p proposed by the principal, the agent computes his value of μ that maximizes his expected utility rate and subsequently decides whether to accept the contract or not, by solving the following optimization problem:

$$\max_{\mu \geq 0} u(\mu) = \max_{\mu \geq 0} \left\{ w - \frac{p\lambda}{\lambda + \mu} - \mu - \eta p \left(\frac{1}{2} - \left| \frac{1}{2} - \frac{\lambda}{\lambda + \mu} \right| \right) \right\} \tag{4.4}$$

The agent's optimal service capacity is denoted by $\mu^*(w, p) = \operatorname{argmax}_{\mu \geq 0} u(\mu)$.

Notation:

$$p_1 \equiv \frac{\lambda}{1 + \eta}, \; p_2 \equiv \frac{\lambda}{1 - \eta}, \text{ and } p_3 \equiv \frac{8\left(1 - \sqrt{1 - \eta^2}\right)\lambda}{\eta^2} \tag{4.5}$$

and the following identity is easily verified using the definition of p_3:

$$w_3 \equiv \eta p_3 + 2\sqrt{(1 - \eta)p_3\lambda} - \lambda = 2\sqrt{(1 + \eta)p_3\lambda} - \lambda \tag{4.6}$$

p_1, p_2, p_3 and w_3 are functions of λ and η. However we suppress the parameters (λ, η).

Next we state a number of technical lemmas (see proofs in the Appendix).

Lemma 4.1. *Let* $1 > \eta > 0$ *and* $\lambda > 0$. *If* $p \geq \lambda/(1 - \eta)$, *then* $p \geq \eta p + 2\sqrt{(1 - \eta)p\lambda} - \lambda > 0$.

Lemma 4.2. *Let* $1 > \eta > 0$ *and* $\lambda > 0$.

(a) *If* $p > 8\left(1 - \sqrt{1 - \eta^2}\right)\lambda/\eta^2$, *then* $\eta p - 2\left(\sqrt{1 + \eta} - \sqrt{1 - \eta}\right)\sqrt{p\lambda} > 0$.

(b) *If* $8\left(1 - \sqrt{1 - \eta^2}\right)\lambda/\eta^2 > p > 0$, *then* $0 > \eta p - 2\left(\sqrt{1 + \eta} - \sqrt{1 - \eta}\right)\sqrt{p\lambda}$.

(c) *If* $p = 8\left(1-\sqrt{1-\eta^2}\right)\lambda/\eta^2$, *then* $\eta p - 2\left(\sqrt{1+\eta}-\sqrt{1-\eta}\right)\sqrt{p\lambda} = 0$.

Lemma 4.3. *Let* $1 > \eta > 0$ *and* $\lambda > 0$, *then* $4\lambda/(1-\eta) > 8\left(1-\sqrt{1-\eta^2}\right)\lambda/\eta^2 > 4\lambda/(1+\eta)$.

Lemma 4.4. *Let* $\eta > 0$ *and* $\lambda > 0$. *If* $p > 4\lambda/(1+\eta)$, *then* $2\sqrt{(1+\eta)p\lambda} - \lambda > 0$.

Lemma 4.5. *Let* $\eta > 0$ *and* $\lambda > 0$.

(a) *If* $\left(1+2\eta+2\sqrt{\eta(1+\eta)}\right)\lambda > p > \left(1+2\eta-2\sqrt{\eta(1+\eta)}\right)\lambda$, *then* $0 > p - 2\sqrt{(1+\eta)p\lambda} + \lambda$.
(b) *If* $\left(1+2\eta-2\sqrt{\eta(1+\eta)}\right)\lambda > p > 0$ *or* $p > \left(1+2\eta+2\sqrt{\eta(1+\eta)}\right)\lambda$, *then* $p - 2\sqrt{(1+\eta)p\lambda} + \lambda > 0$.
(c) *If* $p = \left(1+2\eta-2\sqrt{\eta(1+\eta)}\right)\lambda$ *or* $\left(1+2\eta+2\sqrt{\eta(1+\eta)}\right)\lambda$, *then* $p - 2\sqrt{(1+\eta)p\lambda} + \lambda = 0$.

Lemma 4.6. *Let* $\eta > 0$ *and* $\lambda > 0$, *then* $4\lambda/(1+\eta) > \left(1+2\eta-2\sqrt{\eta(1+\eta)}\right)\lambda$.

Lemma 4.7. *Let* $\lambda > 0$.

(a) *If* $4/5 > \eta > 0$, *then* $\left(1+2\eta+2\sqrt{\eta(1+\eta)}\right)\lambda > \lambda/(1-\eta)$.
(b) *If* $1 > \eta > 4/5$, *then* $\lambda/(1-\eta) > \left(1+2\eta+2\sqrt{\eta(1+\eta)}\right)$.
(c) *If* $\eta = 4/5$, *then* $\left(1+2\eta+2\sqrt{\eta(1+\eta)}\right)\lambda = \lambda/(1-\eta)$.

Lemma 4.8. *Let* $\lambda > 0$.

(a) *If* $4/5 > \eta > 0$, *then* $8\left(1-\sqrt{1-\eta^2}\right)\lambda/\eta^2 > \lambda/(1-\eta)$.
(b) *If* $1 > \eta > 4/5$, *then* $\lambda/(1-\eta) > 8\left(1-\sqrt{1-\eta^2}\right)\lambda/\eta^2$.
(c) *If* $\eta = 4/5$, *then* $8\left(1-\sqrt{1-\eta^2}\right)\lambda/\eta^2 = \lambda/(1-\eta)$.

Lemma 4.8 part (a) implies $\eta \in (0, 4/5) \Rightarrow p_3 > p_2$, which makes condition (4.8) below consistent.

We identify the optimal response of a weakly risk-averse agent to any contract offer $(w, p) \in \mathbb{R}^2_+$ in Proposition 4.9.

Proposition 4.9. *Consider a weakly risk-averse agent* ($\eta \in (0, 4/5)$).

(a) *Given*

$$w \geq p \in (0, p_2] \tag{4.7}$$

then the agent accepts the contract and installs $\mu^*(w,p) = 0$ *resulting in expected utility rate* $u_A(\mu^*(w,p); w, p) = w - p \geq 0$. *The agent rejects the contract if* $p \in (0, p_2]$ *and* $w \in (0, p)$.

(b) *Given*

$$p \in (p_2, p_3) \text{ and } w \geq \eta p + 2\sqrt{(1-\eta)p\lambda} - \lambda \tag{4.8}$$

then the agent accepts the contract and installs $\mu^*(w,p) = \sqrt{(1-\eta)p\lambda} - \lambda > 0$ *resulting in expected utility rate* $u_A(\mu^*(w,p); w,p) = w - \eta p - 2\sqrt{(1-\eta)p\lambda} + \lambda \geq 0$. *The agent rejects the contract if* $p \in (p_2, p_3)$ *and* $w \in \left(0, \eta p + 2\sqrt{(1-\eta)p\lambda} - \lambda\right)$.

(c) *Given*

$$p = p_3 \text{ and } w \geq w_3 \tag{4.9}$$

then the agent accepts the contract and is indifferent about installing either $\mu^*(w,p) = \sqrt{(1-\eta)p_3\lambda} - \lambda$ *or* $\mu^*(w,p) = \sqrt{(1+\eta)p_3\lambda} - \lambda$. *In both cases the agent receives expected utility rate* $u_A(\mu^*(w,p); w,p) = w - w_3 \geq 0$. *If* $r \in (0, p_3)$, *then there exists a* w^* *such that* $\left((w^*, p_3), \sqrt{(1-\eta)p_3\lambda} - \lambda\right)$ *is the unique admissible solution (see Definition 2.3). If* $r = p_3$, *then there exists* w^* *such that* $\left((w^*, p_3), \sqrt{(1-\eta)p_3\lambda} - \lambda\right)$ *and* $\left((w^*, p_3), \sqrt{(1+\eta)p_3\lambda} - \lambda\right)$ *are both admissible solutions (see Definition 2.3). If* $r > p_3$, *then there exists a* w^* *such that* $\left((w^*, p_3), \sqrt{(1+\eta)p_3\lambda} - \lambda\right)$ *is the unique admissible solution (for proof see Proposition 4.12). He rejects the contract if* $p = p_3$ *and* $w \in (0, w_3)$.

(d) *Given*

$$p > p_3 \text{ and } w \geq 2\sqrt{(1+\eta)p\lambda} - \lambda \tag{4.10}$$

then the agent accepts the contract and installs $\mu^*(w,p) = \sqrt{(1+\eta)p\lambda} - \lambda > 0$ *resulting in expected utility rate* $u_A(\mu^*(w,p); w,p) = w - 2\sqrt{(1+\eta)p\lambda} + \lambda \geq 0$. *The agent rejects the contract if* $p > p_3$ *and* $w \in \left(0, 2\sqrt{(1+\eta)p\lambda} - \lambda\right)$.

Proof. According to Table 4.1, the optimization of $u(\mu)$ when $\eta \in (0, 3/5]$ versus $\eta \in (3/5, 4/5)$ is different. Therefore we prove the proposition separately for $\eta \in (0, 3/5]$ and $\eta \in (3/5, 4/5)$.

Case $\eta \in (0, 3/5]$: Note that $4p_2 > 4p_1 \geq p_2$ and according to Lemma 4.3, $4p_2 > p_3 > 4p_1$. Therefore we have $4p_2 > p_3 > 4p_1 \geq p_2$. Figure 4.2 shows the shape of $u(\mu)$ when $\eta \in (0, 3/5]$ and the value of p falls in different ranges. The structure of the proof when $\eta \in (0, 3/5]$ is depicted in Fig. 4.3.

Case $p \in (0, p_2]$: According to Table 4.1, $u(\mu)$ is decreasing with respect to $\mu \geq 0$. Thus the agent's optimal service capacity is $\mu^*(w,p) = 0$ and from (4.3) $u(\mu^*(w,p)) = w - p$.

Subcase $w \in (0, p)$: $u(\mu^*(w,p)) < 0$, therefore the agent rejects the contract.

Subcase $w \geq p$: $u(\mu^*(w,p)) \geq 0$, thus the agent would accept the contract if offered.

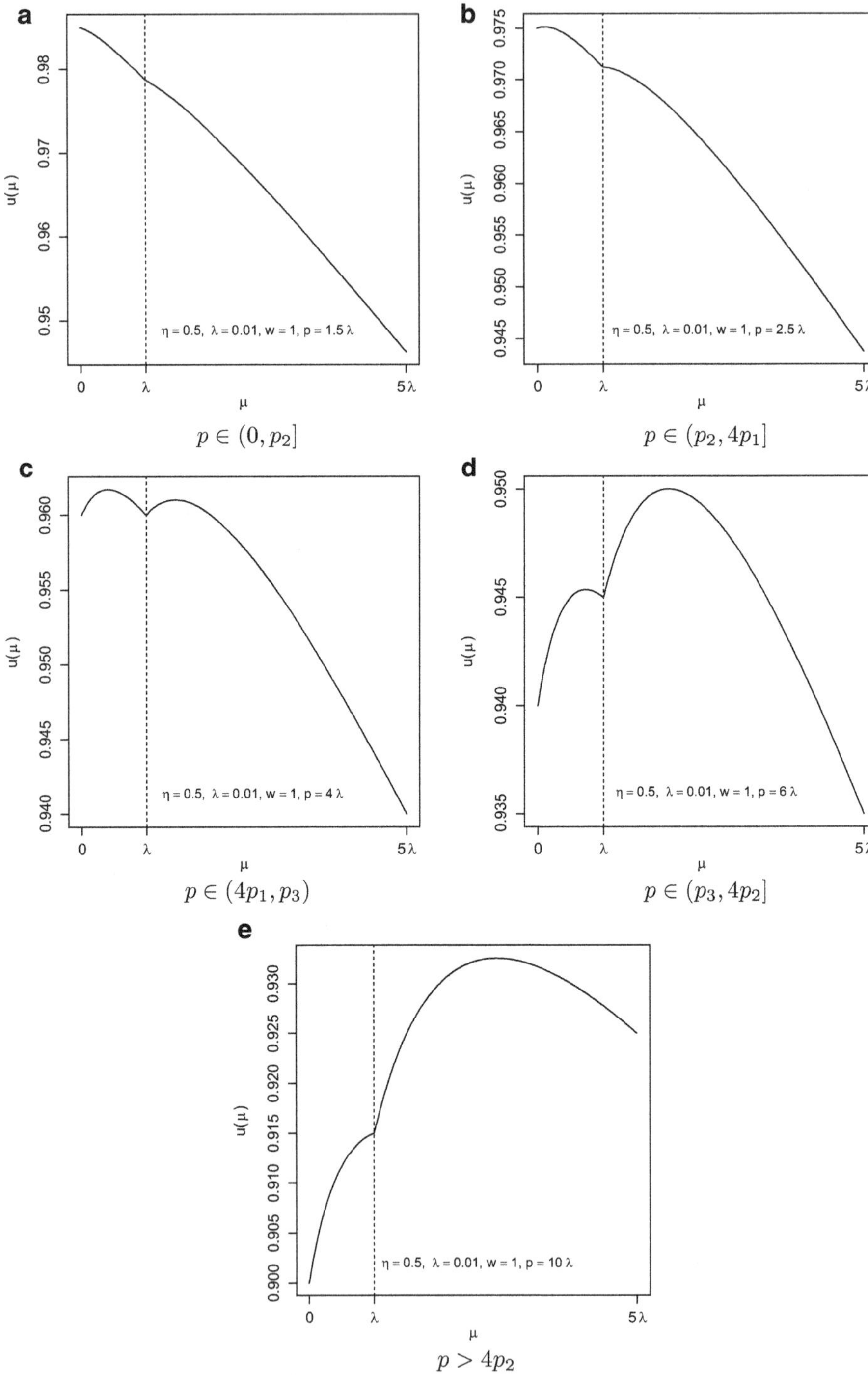

Fig. 4.2 Illustration of the forms of $u(\mu)$ when $\eta \in (0, 3/5]$

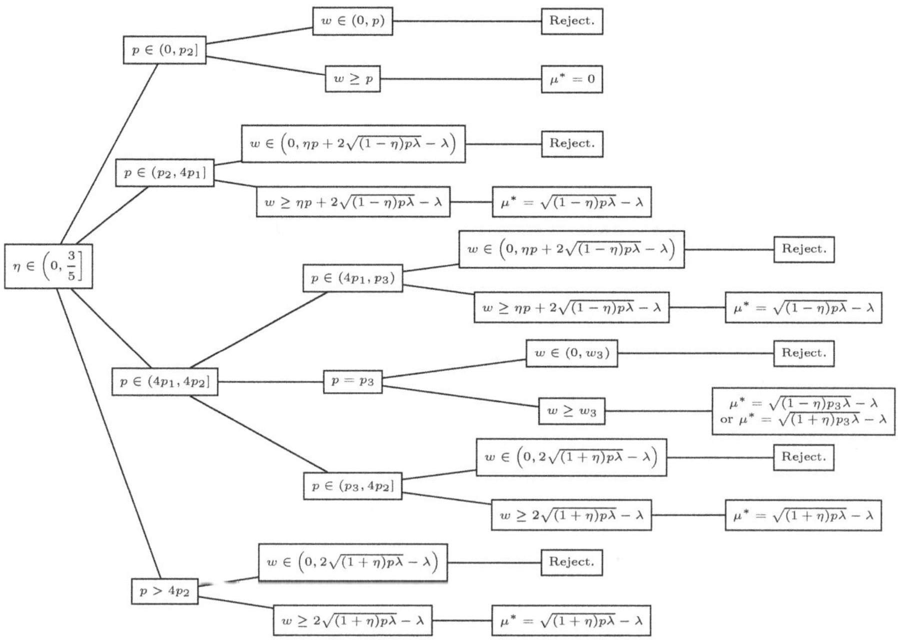

Fig. 4.3 Structure of the proof for Proposition 4.9 when $\eta \in (0, 3/5]$

Case $p \in (p_2, 4p_1]$: According to Table 4.1, the service capacity that maximizes $u(\mu)$ lies in $(0, \lambda)$. $\mu^*(w, p)$ is computed from first order condition $du(\mu)/d\mu|_{\mu=\mu^*(w,p)} = 0 \Rightarrow \mu^*(w, p) = \sqrt{(1-\eta)p\lambda} - \lambda > 0$ and from Eq. (4.3) $u(\mu^*(w, p)) = w - \eta p - 2\sqrt{(1-\eta)p\lambda} + \lambda$. According to Lemma 4.1, $p > p_2 \Rightarrow \eta p + 2\sqrt{(1-\eta)p\lambda} - \lambda > 0$.

Subcase $w \in \left(0, \eta p + 2\sqrt{(1-\eta)p\lambda} - \lambda\right)$: $u(\mu^*(w, p)) < 0$, therefore the agent rejects the contract.

Subcase $w \geq \eta p + 2\sqrt{(1-\eta)p\lambda} - \lambda$: $u(\mu^*(w, p)) \geq 0$, therefore the agent would accept the contract if offered.

Case $p \in (4p_1, 4p_2]$: According to Table 4.1, there is a service capacity that maximizes $u(\mu)$ for $\mu \in (0, \lambda]$ and a service capacity that maximizes $u(\mu)$ for $\mu > \lambda$. Denote the optimal service capacity in $(0, \lambda]$ by $\mu^*_{(0,\lambda]}(w, p)$. From the first order condition the optimal service capacity is $\mu^*_{(0,\lambda]}(w, p) = \sqrt{(1-\eta)p\lambda} - \lambda$ and from (4.3) $u\left(\mu^*_{(0,\lambda]}(w, p)\right) = w - \eta p - 2\sqrt{(1-\eta)p\lambda} + \lambda$. Denote the optimal service capacity for $\mu > \lambda$ by $\mu^*_\lambda(w, p)$, which is solved from first order condition $du(\mu)/d\mu|_{\mu=\mu^*_\lambda(w,p)} = 0 \Rightarrow \mu^*_\lambda(w, p) = \sqrt{(1+\eta)p\lambda} - \lambda$ and from (4.3) $u\left(\mu^*_\lambda(w, p)\right) = w - 2\sqrt{(1+\eta)p\lambda} + \lambda$. The agent has a choice of two service capacities and he installs the one that generates a higher expected utility rate.

Note that $u\left(\mu^*_{\lambda}(w,p)\right) - u\left(\mu^*_{(0,\lambda]}(w,p)\right) = \eta p - 2\left(\sqrt{1+\eta} - \sqrt{1-\eta}\right)\sqrt{p\lambda}$. According to Lemma 4.3, $4p_2 > p_3 > 4p_1$, therefore we examine the following subcases.

Subcase $p \in (4p_1, p_3)$: From Lemma 4.2 part (b), $u\left(\mu^*_{(0,\lambda]}(w,p)\right) > u\left(\mu^*_{\lambda}(w,p)\right)$, thus the agent's optimal service capacity is $\mu^*(w,p) = \sqrt{(1-\eta)p\lambda} - \lambda$ and $u(\mu^*(w,p)) = w - \eta p - 2\sqrt{(1-\eta)p\lambda} + \lambda$. From Lemma 4.1, $p > 4p_1 \geq p_2 \Rightarrow \eta p + 2\sqrt{(1-\eta)p\lambda} - \lambda > 0$.

Subsubcase $w \in \left(0, \eta p + 2\sqrt{(1-\eta)p\lambda} - \lambda\right)$: $u(\mu^*(w,p)) < 0$, therefore the agent rejects the contract.

Subsubcase $w \geq \eta p + 2\sqrt{(1-\eta)p\lambda} - \lambda$: $u(\mu^*(w,p)) \geq 0$, thus the agent would accept the contract if offered.

Subcase $p = p_3$: According to Lemma 4.2 part (c), $u\left(\mu^*_{(0,\lambda]}(w,p_3)\right) = u\left(\mu^*_{\lambda}(w,p_3)\right)$, indicating that installing $\mu^*_{(0,\lambda]}(w,p_3)$ or $\mu^*_{\lambda}(w,p_3)$ leads to the same agent's expected utility rate. Therefore the agent is indifferent about installing either $\mu^*(w,p) = \sqrt{(1-\eta)p_3\lambda} - \lambda$ or $\mu^*(w,p) = \sqrt{(1+\eta)p_3\lambda} - \lambda$. Still, the capacity value has to lead to admissible solutions (see Proposition 4.12). Recall the definition of w_3 from (4.6). According to Lemma 4.1, $p_3 > 4p_1 \geq p_2 \Rightarrow w_3 = \eta p_3 + 2\sqrt{(1-\eta)p_3\lambda} - \lambda > 0$.

Subsubcase $w \in (0, w_3)$: $u(\mu^*(w,p)) < 0$, thus the agent rejects the contract.

Subsubcase $w \geq w_3$: $u(\mu^*(w,p)) \geq 0$, thus the agent would accept the contract if offered.

Subcase $p \in (p_3, 4p_2]$: By Lemma 4.2 part (a), $u\left(\mu^*_{\lambda}(w,p)\right) > u\left(\mu^*_{(0,\lambda]}(w,p)\right)$, therefore the agent's optimal service capacity is $\mu^*(w,p) = \sqrt{(1+\eta)p\lambda} - \lambda$ and $u(\mu^*(w,p)) = w - 2\sqrt{(1+\eta)p\lambda} + \lambda$. According to Lemma 4.4, $p > p_3 > 4p_1 \Rightarrow 2\sqrt{(1+\eta)p\lambda} - \lambda > 0$.

Subsubcase $w \in \left(0, 2\sqrt{(1+\eta)p\lambda} - \lambda\right)$: $u(\mu^*(w,p)) < 0$, therefore the agent rejects the contract.

Subsubcase $w \geq 2\sqrt{(1+\eta)p\lambda} - \lambda$: $u(\mu^*(w,p)) \geq 0$, therefore the agent would accept the contract if offered.

Case $p > 4p_2$: According to Table 4.1, the service capacity that maximizes $u(\mu)$ satisfies $\mu > \lambda$. From the first order condition the agent's optimal service capacity is $\mu^*(w,p) = \sqrt{(1+\eta)p\lambda} - \lambda$ and from Eq. (4.3) $u(\mu^*(w,p)) = w - 2\sqrt{(1+\eta)p\lambda} + \lambda$. According to Lemma 4.4, $p > 4p_2 > 4p_1 \Rightarrow 2\sqrt{(1+\eta)p\lambda} - \lambda > 0$.

Subcase $w \in \left(0, 2\sqrt{(1+\eta)p\lambda} - \lambda\right)$: $u(\mu^*(w,p)) < 0$, thus the agent rejects the contract.

Subcase $w \geq 2\sqrt{(1+\eta)p\lambda} - \lambda$: $u(\mu^*(w,p)) \geq 0$, therefore the agent would accept the contract if offered.

This completes the proof for Proposition 4.9 when $\eta \in (0, 3/5]$.

Case $\eta \in (3/5, 4/5)$: Note that $4p_2 > p_2 > 4p_1$ and according to Lemmas 4.3 and 4.8 part (a), $4p_2 > p_3 > p_2$. Therefore we have $4p_2 > p_3 > p_2 > 4p_1$. Figure 4.4 shows the shape of $u(\mu)$ when $\eta \in (3/5, 4/5)$ and the value of p falls in different ranges. The structure of the proof when $\eta \in (3/5, 4/5)$ is depicted in Fig. 4.5.

Case $p \in (0, 4p_1]$: According to Table 4.1, $u(\mu)$ is decreasing with respect to $\mu \geq 0$. Thus the agent's optimal service capacity is $\mu^*(w, p) = 0$ and from (4.3) $u(\mu^*(w, p)) = w - p$.

Subcase $w \in (0, p)$: $u(\mu^*(w, p)) < 0$, therefore the agent rejects the contract.
Subcase $w \geq p$: $u(\mu^*(w, p)) \geq 0$, thus the agent would accept the contract if offered.

Case $p \in (4p_1, p_2]$: According to Table 4.1, there is a service capacity that maximizes $u(\mu)$ for $\mu \in [0, \lambda)$ and a service capacity that maximizes $u(\mu)$ for $\mu > \lambda$. Denote the optimal service capacity in $[0, \lambda)$ by $\mu^*_{[0,\lambda)}(w, p)$. Note that $u(\mu)$ is decreasing with respect to μ over $[0, \lambda)$, therefore the agent's optimal service capacity is $\mu^*_{[0,\lambda)}(w, p) = 0$ and from (4.3) $u\left(\mu^*_{[0,\lambda)}(w, p)\right) = w - p$. Denote the optimal service capacity for $\mu > \lambda$ by $\mu^*_\lambda(w, p)$. From the first order condition $\mu^*_\lambda(w, p) = \sqrt{(1 + \eta)p\lambda} - \lambda$ and from Eq. (4.3) $u\left(\mu^*_\lambda(w, p)\right) = w - 2\sqrt{(1 + \eta)p\lambda} + \lambda$. The agent has to choose one of the two service capacities and installs the one with higher expected utility rate. Note that $u\left(\mu^*_\lambda(w, p)\right) - u\left(\mu^*_{[0,\lambda)}(w, p)\right) = p - 2\sqrt{(1 + \eta)p\lambda} + \lambda$. According to Lemma 4.6, $4p_1 > \left(1 + 2\eta - 2\sqrt{\eta(1 + \eta)}\right)\lambda$ and according to Lemma 4.7 part (a), $\left(1 + 2\eta + 2\sqrt{\eta(1 + \eta)}\right)\lambda > p_2$. Thus according to Lemma 4.5 part (a), $u\left(\mu^*_{[0,\lambda)}(w, p)\right) > u\left(\mu^*_\lambda(w, p)\right)$, the agent's optimal service capacity is $\mu^*(w, p) = 0$ and $u(\mu^*(w, p)) = w - p$.

Subcase $w \in (0, p)$: $u(\mu^*(w, p)) < 0$, therefore the agent rejects the contract.
Subcase $w \geq p$: $u(\mu^*(w, p)) \geq 0$, thus the agent would accept the contract if offered.

Case $p \in (p_2, 4p_2]$: According to Table 4.1, there is a service capacity that maximizes $u(\mu)$ for $\mu \in (0, \lambda]$ and a service capacity that maximizes $u(\mu)$ for $\mu > \lambda$. Denote the optimal service capacity in $(0, \lambda]$ by $\mu^*_{(0,\lambda]}(w, p)$. From the first order condition $\mu^*_{(0,\lambda]}(w, p) = \sqrt{(1 - \eta)p\lambda} - \lambda$ and from Eq. (4.3) $u\left(\mu^*_{(0,\lambda]}(w, p)\right) = w - \eta p - 2\sqrt{(1 - \eta)p\lambda} + \lambda$. Denote the optimal service capacity for $\mu > \lambda$ by $\mu^*_\lambda(w, p)$. From the first order condition $\mu^*_\lambda(w, p) = \sqrt{(1 + \eta)p\lambda} - \lambda$ and from Eq. (4.3) $u\left(\mu^*_\lambda(w, p)\right) = w - 2\sqrt{(1 + \eta)p\lambda} + \lambda$. The agent has to choose one of the two service capacities and installs the one that generates a higher expected utility rate. Note that $u\left(\mu^*_\lambda(w, p)\right) - u$

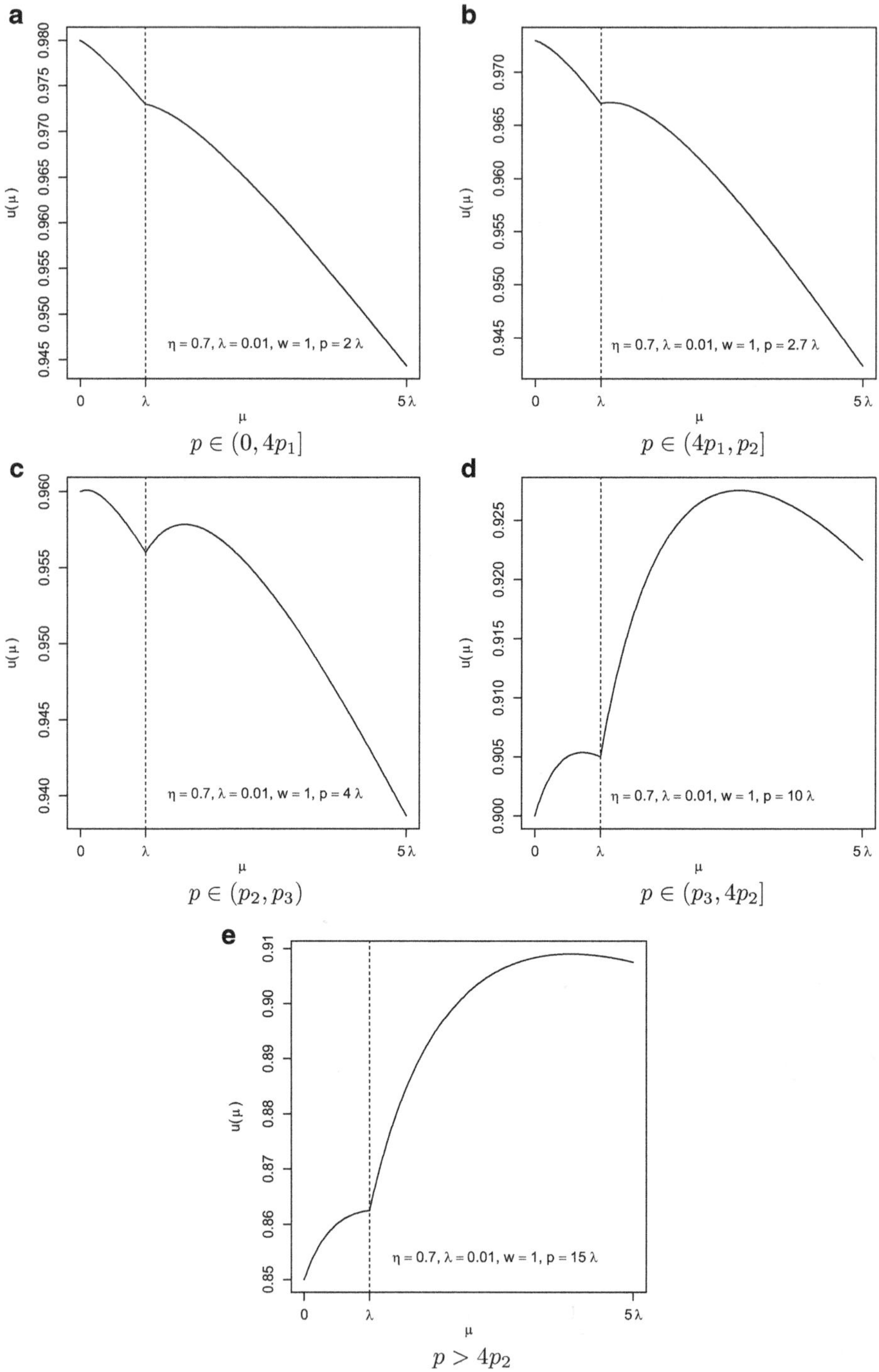

Fig. 4.4 Illustration of the forms of $u(\mu)$ when $\eta \in (3/5, 4/5)$

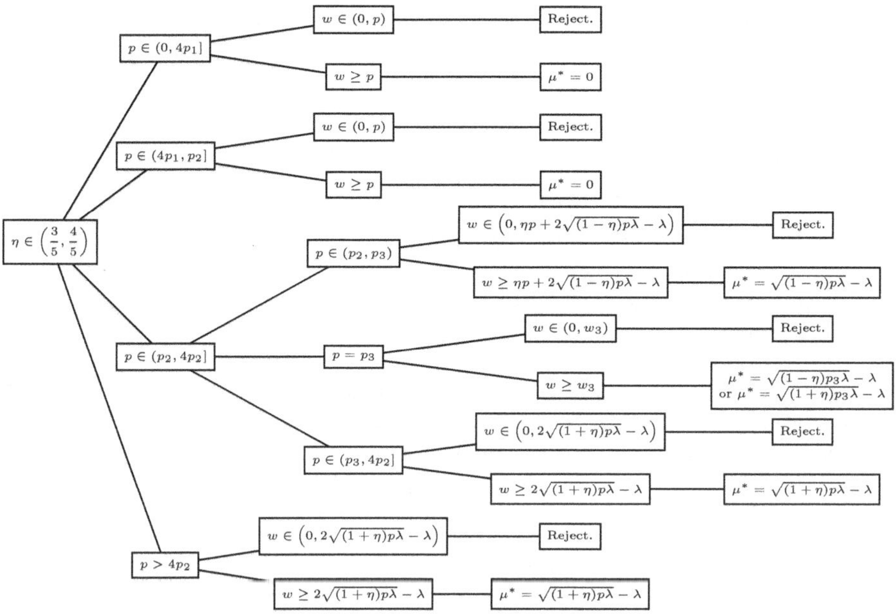

Fig. 4.5 Structure of the proof for Proposition 4.9 when $\eta \in (3/5, 4/5)$

$\left(\mu^*_{(0,\lambda]}(w,p)\right) = \eta p - 2\left(\sqrt{1+\eta} - \sqrt{1-\eta}\right)\sqrt{p\lambda}$. According to Lemmas 4.3 and 4.8 part (a), $4p_2 > p_3 > p_2$, therefore we examine the following subcases.

Subcase $p \in (p_2, p_3)$: By Lemma 4.2 part (b), $u\left(\mu^*_{(0,\lambda]}(w,p)\right) > u\left(\mu^*_{\lambda}(w,p)\right)$, therefore the agent's optimal service capacity is $\mu^*(w,p) = \sqrt{(1-\eta)p\lambda} - \lambda$ and $u(\mu^*(w,p)) = w - \eta p - 2\sqrt{(1-\eta)p\lambda} + \lambda$. According to Lemma 4.1, $p > p_2 \Rightarrow \eta p + 2\sqrt{(1-\eta)p\lambda} - \lambda > 0$.

Subsubcase $w \in \left(0, \eta p + 2\sqrt{(1-\eta)p\lambda} - \lambda\right)$: $u(\mu^*(w,p)) < 0$, therefore the agent rejects the contract.

Subsubcase $w \geq \eta p + 2\sqrt{(1-\eta)p\lambda} - \lambda$: $u(\mu^*(w,p)) \geq 0$, thus the agent would accept the contract if offered.

Subcase $p = p_3$: According to Lemma 4.2 part (c), $u\left(\mu^*_{(0,\lambda]}(w,p_3)\right) = u\left(\mu^*_{\lambda}(w,p_3)\right)$, indicating that installing $\mu^*_{(0,\lambda]}(w,p)$ or $\mu^*_{\lambda}(w,p)$ leads to the same agent's expected utility rate. Therefore the agent is indifferent about installing $\mu^*(w,p) = \sqrt{(1-\eta)p_3\lambda} - \lambda$ or $\mu^*(w,p) = \sqrt{(1+\eta)p_3\lambda} - \lambda$. Again, the service capacity has to lead to admissible solutions (see Proposition 4.12). Recall the definition of w_3 from (4.6). According to Lemma 4.1, $p_3 > p_2 \Rightarrow w_3 = \eta p_3 + 2\sqrt{(1-\eta)p_3\lambda} - \lambda > 0$.

Subsubcase $w \in (0, w_3)$: $u(\mu^*(w,p)) < 0$, thus the agent rejects the contract.

Subsubcase $w \geq w_3$: $u(\mu^*(w,p)) \geq 0$, thus the agent would accept the contract if offered.

Subcase $p \in (p_3, 4p_2]$: By Lemma 4.2 part (a), $u\left(\mu^*_\lambda(w,p)\right) > u\left(\mu^*_{(0,\lambda]}(w,p)\right)$, thus the agent's optimal service capacity is $\mu^*(w,p) = \sqrt{(1+\eta)p\lambda} - \lambda$ and $u(\mu^*(w,p)) = w - 2\sqrt{(1+\eta)p\lambda} + \lambda$. According to Lemma 4.4, $p > p_3 > p_2 > 4p_1 \Rightarrow 2\sqrt{(1+\eta)p\lambda} - \lambda > 0$.

Subsubcase $w \in \left(0, 2\sqrt{(1+\eta)p\lambda} - \lambda\right)$: $u(\mu^*(w,p)) < 0$, therefore the agent rejects the contract.

Subsubcase $w \geq 2\sqrt{(1+\eta)p\lambda} - \lambda$: $u(\mu^*(w,p)) \geq 0$, therefore the agent would accept the contract if offered.

Case $p > 4p_2$: According to Table 4.1, the service capacity that maximizes $u(\mu)$ satisfies $\mu > \lambda$. From the first order condition the agent's optimal service capacity is $\mu^*(w,p) = \sqrt{(1+\eta)p\lambda} - \lambda$ and from Eq. (4.3) $u(\mu^*(w,p)) = w - 2\sqrt{(1+\eta)p\lambda} + \lambda$. According to Lemma 4.4, $p > 4p_2 > 4p_1 \Rightarrow 2\sqrt{(1+\eta)p\lambda} - \lambda > 0$.

Subcase $w \in \left(0, 2\sqrt{(1+\eta)p\lambda} - \lambda\right)$: $u(\mu^*(w,p)) < 0$, thus the agent rejects the contract.

Subcase $w \geq 2\sqrt{(1+\eta)p\lambda} - \lambda$: $u(\mu^*(w,p)) \geq 0$, therefore the agent would accept the contract if offered.

This completes the proof for Proposition 4.9 when $\eta \in (3/5, 4/5)$. □

In summary, given exogenous market conditions such that a contract offer satisfying the reservation value constraints for both the principal and a weakly risk-averse agent exists (see Theorem 4.19 and Proposition 4.20 later), the agent determines his optimal capacity using one of two formulas:

$$\mu^*(w,p) = \sqrt{(1-\eta)p\lambda} - \lambda > 0 \text{ or } \mu^*(w,p) = \sqrt{(1+\eta)p\lambda} - \lambda > 0$$

The conditions when a weakly risk-averse agent accepts the contract can be depicted by the shaded areas in Fig. 4.6, where $\eta = 0.6$. The three shaded areas with different grey scales represent conditions (4.7), (4.8) and (4.10) under which the agent accepts the contract but responds differently. The lower bound function of the shaded areas (denoted by $w_0(p)$) represents the set of offers with agent's zero expected utility rate. $w_0(p)$ is defined as follows:

$$w_0(p) = \begin{cases} p \text{ when } p \in (0, p_2] \\ \eta p + 2\sqrt{(1-\eta)p\lambda} - \lambda \text{ when } p \in (p_2, p_3] \\ 2\sqrt{(1+\eta)p\lambda} - \lambda \text{ when } p > p_3 \end{cases}$$

Note that since $\lim_{p\to p_2^-} w_0(p) = \lim_{p\to p_2^+} w_0(p) = p_2$ and $\lim_{p\to p_3^-} w_0(p) = \lim_{p\to p_3^+} = \eta p_3 + 2\sqrt{(1-\eta)p_3\lambda} - \lambda$, $w_0(p)$ is continuous everywhere over interval

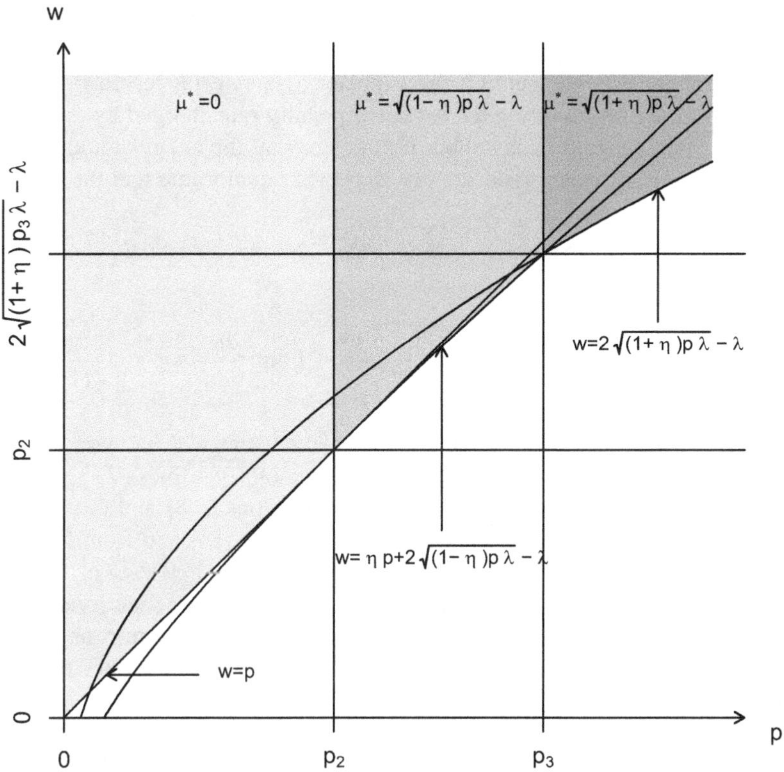

Fig. 4.6 Conditions when a weakly risk-averse agent accepts the contract with $\eta = 0.6$

$p \in \mathbb{R}_+$. Since $\lim_{p\to p_2^-} dw_0(p)/dp = \lim_{p\to p_2^+} dw_0(p)/dp = 1$, $w_0(p)$ is differentiable at $p = p_2$. However since $\lim_{p\to p_3^-} dw_0(p)/dp = \eta + \sqrt{(1-\eta)\lambda/p_3} \neq \sqrt{(1+\eta)\lambda/p_3} = \lim_{p\to p_3^+} dw_0(p)/dp$, $w_0(p)$ is not differentiable at $p = p_3$.

4.1.1 *Sensitivity Analysis of a Weakly Risk-Averse Agent's Optimal Strategy*

The principal would not propose an acceptable contract that results in $u_A(\mu^* = 0) \geq \underline{u_A} = 0$. Therefore the only viable cases to consider are when the agent accepts the contract and installs positive service capacities: $\mu^*(w,p) = \sqrt{(1-\eta)p\lambda} - \lambda$ or $\mu^*(w,p) = \sqrt{(1+\eta)p\lambda} - \lambda$. We examine the two viable contracts with positive service capacities.

First the case $\mu^*(w,p) = \sqrt{(1-\eta)p\lambda} - \lambda$. According to (4.8) the compensation rate w is bounded below by $\eta p + 2\sqrt{(1-\eta)p\lambda} - \lambda = \eta pP(0) + pP(1) + \mu^*(w,p)$, with the term $\eta pP(0)$ representing the expected risk rate perceived by the agent and the term $pP(1)$ representing the expected penalty rate charged by the principal when the optimal capacity is installed. It indicates that the agent should at least be reimbursed for the expected risk rate, the expected penalty rate and the cost of the optimal service capacity.

The optimal service capacity $\sqrt{(1-\eta)p\lambda} - \lambda$ depends on p, λ, and η. Its derivatives are:

$$\frac{\partial \mu^*}{\partial p} = \sqrt{\frac{(1-\eta)\lambda}{4p}} > 0, \ \frac{\partial \mu^*}{\partial \lambda} = \sqrt{\frac{(1-\eta)p}{4\lambda}} - 1 \text{ and } \frac{\partial \mu^*}{\partial \eta} = -\sqrt{\frac{p\lambda}{4(1-\eta)}} < 0$$

These derivatives indicate that given a λ and η, the agent will increase his service capacity when the penalty rate increases. Note that $\sqrt{(1-\eta)p\lambda} - \lambda$, as a function of λ, decreases when $\lambda > (1-\eta)p/4$. From conditions (4.8) and (4.9) the agent installs service capacity $\sqrt{(1-\eta)p\lambda} - \lambda$ when $p \in (p_2, p_3]$ and from Lemma 4.3 we have $4p_2 > p_3$. Therefore we have $4\lambda/(1-\eta) = 4p_2 > p \Rightarrow \lambda > (1-\eta)p/4 \Rightarrow \partial\mu^*/\partial\lambda < 0$. Thus, given a p and η, the savings from reducing the service capacity are greater than the increase in the penalty charge and in the risk rate, and the agent will reduce μ when λ increases. Given a p and λ, the agent will reduce the μ when he is more risk-averse.

The agent's optimal expected utility rate when installing capacity $\mu^*(w,p) = \sqrt{(1-\eta)p\lambda} - \lambda$ is $u_A^* \equiv u_A(\mu^*(w,p); w,p) = w - \eta p - 2\sqrt{(1-\eta)p\lambda} + \lambda$, and it depends on w, p, η and λ. Note that $\partial u_A^*/\partial w = -1 < 0$, $\partial u_A^*/\partial p = -\eta - \sqrt{(1-\eta)\lambda/p} < 0$, indicating that the agent's optimal expected utility rate decreases with the compensation rate and the penalty rate. Note that $\partial u_A^*/\partial \eta = -\sqrt{p}\left(\sqrt{p} - \sqrt{p_2}\right)$ and $\partial u_A^*/\partial \lambda = -\left(\sqrt{p} - \sqrt{p_2}\right)/\sqrt{p_2}$, and from Proposition 4.9 $p > p_2 \Rightarrow \sqrt{p} - \sqrt{p_2} > 0$, therefore the agent's optimal expected utility rate also decreases with his risk intensity and the failure rate.

Next we examine the case $\mu^*(w,p) = \sqrt{(1+\eta)p\lambda} - \lambda$. According to (4.10) the compensation rate w is bounded below by $2\sqrt{(1+\eta)p\lambda} - \lambda = \eta pP(1) + pP(1) + \mu^*(w,p)$, with the term $\eta pP(1)$ representing the expected risk rate perceived by the agent and $pP(1)$ representing the expected penalty rate charged by the principal when the optimal capacity is installed. It indicates that the agent should at least be reimbursed for the expected risk rate, the expected penalty rate and the cost of the optimal service capacity.

The optimal service capacity $\sqrt{(1+\eta)p\lambda} - \lambda$ depends on p, λ, and η. Its derivatives are:

$$\frac{\partial \mu^*}{\partial p} = \sqrt{\frac{(1+\eta)\lambda}{4p}} > 0, \ \frac{\partial \mu^*}{\partial \lambda} = \sqrt{\frac{(1+\eta)p}{4\lambda}} - 1 \text{ and } \frac{\partial \mu^*}{\partial \eta} = \sqrt{\frac{p\lambda}{4(1+\eta)}} > 0$$

The derivatives indicate that given λ and η, the agent will increase the μ when the penalty rate increases. Note that $\sqrt{(1+\eta)p\lambda} - \lambda$, as a function of λ, increases when $(1+\eta)p/4 > \lambda$. From (4.9) and (4.10) the agent installs service capacity $\sqrt{(1+\eta)p\lambda} - \lambda$ when $p \geq p_3$, and from Lemma 4.3 we have $p_3 > 4p_1$. Therefore we have $p > 4p_1 = 4\lambda/(1+\eta) \Rightarrow (1+\eta)p/4 > \lambda \Rightarrow \partial\mu^*/\partial\lambda > 0$. Thus, given p and η, the agent will increase μ when λ increases. Given p and λ, the agent will increase his μ when he is more risk-averse.

The agent's optimal expected utility rate when installing capacity $\mu^*(w,p) = \sqrt{(1+\eta)p\lambda} - \lambda$ is $u_A^* \equiv u_A(\mu^*(w,p); w,p) = w - 2\sqrt{(1+\eta)p\lambda} + \lambda$, and it depends on w, p, η and λ. Note that $\partial u_A^*/\partial w = -1 < 0$, $\partial u_A^*/\partial p = -\sqrt{(1+\eta)\lambda/p} < 0$ and $\partial u_A^*/\partial\eta = -\sqrt{p\lambda/(1+\eta)} < 0$, indicating that the agent's optimal expected utility rate decreases with the compensation rate, the penalty rate and his risk intensity. Note that $\partial u_A^*/\partial\lambda = -\left(\sqrt{p} - \sqrt{p_1}\right)/\sqrt{p_1}$, and from Proposition 4.9 $p \geq p_3 > p_1 \Rightarrow \sqrt{p} - \sqrt{p_1} > 0$, therefore the agent's optimal expected utility rate also decreases with the failure rate.

Summary: Recall that given the set of offers $\{(w,p) : p \in (0,\lambda], w \geq p\}$ a risk-neutral agent would accept the contract, install $\mu^*(w,p) = 0$ and receive expected utility rate $u(\mu^*(w,p); w,p) = w - p$. Given the set of offers $\left\{(w,p) : p > \lambda, w \geq 2\sqrt{p\lambda} - \lambda\right\}$ he would accept the contract, install $\mu^*(w,p) = \sqrt{p\lambda} - \lambda$ and receive expected utility rate $u(\mu^*(w,p); w,p) = w - 2\sqrt{p\lambda} + \lambda$. By comparing the optimal capacities of a weakly risk-averse agent to that of a risk-neutral agent, three conclusions are drawn:

1. Given a λ, the principal has to set a higher p in order to induce a weakly risk-averse agent to install a positive service capacity versus a risk-neutral agent ($p > \lambda$ for risk-neutral agent, $p > \lambda/(1-\eta)$ for weakly risk-averse agent).
2. Given a λ, when p is relatively low, the μ value plays a more prominent role in the utility of a weakly risk-averse agent who therefore installs a service capacity lower than a risk-neutral agent ($\sqrt{p\lambda} - \lambda > \sqrt{(1-\eta)p\lambda} - \lambda$). As the p increases, the penalty charge and the risk become of greater concern, therefore the weakly risk-averse agent installs a μ^* higher than a risk-neutral agent ($\sqrt{(1+\eta)p\lambda} - \lambda > \sqrt{p\lambda} - \lambda$).
3. In essence, weakly risk-averse attitude makes an agent worse off. We state this conclusion formally in Proposition 4.10.

Proposition 4.10. *Given w and p, an agent who accepts the contract and installs a positive service capacity has a decreasing expected utility rate in $\eta \in [0, 4/5)$.*

Proof. Recall that when w and p satisfy conditions (4.8) and (4.9), the agent installs capacity $\mu^*(w,p) = \sqrt{(1-\eta)p\lambda} - \lambda > 0$, and the agent's expected utility rate is $u\left(\mu^*(w,p)\right) = w - \eta p - 2\sqrt{(1-\eta)p\lambda} + \lambda$. Note that $\partial u/\partial\eta = -p + p\lambda/\sqrt{(1-\eta)p\lambda} = -\left(p - \sqrt{\lambda/(1-\eta)}\sqrt{p}\right) = -\sqrt{p}\left(\sqrt{p} - \sqrt{p_2}\right)$. Since $p > p_2$, therefore $\partial u/\partial\eta < 0$. When the compensation rate w and the penalty rate p satisfy conditions (4.9) and (4.10), the agent installs capacity $\mu^*(w,p) = \sqrt{(1+\eta)p\lambda} - \lambda > 0$, and the agent's expected utility rate is $u(\mu^*(w,p)) = w - 2\sqrt{(1+\eta)p\lambda} + \lambda$, therefore $\partial u/\partial \eta = -\sqrt{p\lambda/(1+\eta)} < 0$. □

The following Corollary follows from Proposition 4.10.

Corollary 4.11. *Given w and p, an agent who accepts the contract and subsequently installs a positive service capacity is always worse off when he is weakly risk-averse ($\eta \in (0, 4/5)$) than risk-neutral ($\eta = 0$).*

We discuss the case for $\eta \geq 4/5$ in Sect. 4.2.1.

4.1.2 Principal's Optimal Strategy

Anticipating the agent's optimal $\mu^*(w,p)$ the principal chooses the w and p that maximize her expected profit rate by solving the optimization problem (4.11).

$$\max_{w>0,p>0} \Pi_P(w,p;\mu^*(w,p)) = \max_{w>0,p>0} \left\{ \frac{r\mu^*(w,p)}{\lambda + \mu^*(w,p)} - w + \frac{p\lambda}{\lambda + \mu^*(w,p)} \right\} \tag{4.11}$$

Denote $(w^*, p^*) = \operatorname{argmax}_{w>0,p>0} \Pi_P(w,p;\mu^*(w,p))$.

Before deriving the principal's optimal strategy, we examine the case when the principal's contract offer satisfies $p = p_3$ and $w \geq w_3$, in which case the agent is indifferent with respect to installing two different service capacities. Nevertheless, the corresponding solutions $((w,p),\mu)$ have to be admissible solutions (see Definition 2.3). We state this case formally in Proposition 4.12.

Proposition 4.12. *Suppose a weakly risk-averse agent. Assume that the principal's potential offers are in the set $\{(w,p) : p = p_3, w \geq w_3\}$.*

(a) *If $r \in (0, p_3)$, the agent installs $\mu^* = \sqrt{(1-\eta)p_3\lambda} - \lambda$ if offered a contract.*
(b) *If $r = p_3$, both $\mu^* = \sqrt{(1-\eta)p_3\lambda} - \lambda$ and $\mu^* = \sqrt{(1+\eta)p_3\lambda} - \lambda$ lead to admissible solutions. Therefore the agent installs either $\sqrt{(1-\eta)p_3\lambda} - \lambda$ or $\sqrt{(1+\eta)p_3\lambda} - \lambda$ if offered a contract.*
(c) *If $r > p_3$, the agent installs $\mu^* = \sqrt{(1+\eta)p_3\lambda} - \lambda$ if offered a contract.*

Proof. Note that for $w \geq w_3$ we have $\partial \Pi_P(w,p_3;\mu)/\partial \mu = (r - p_3)\lambda/(\lambda+\mu)^2$. Define $\mu_L \equiv \sqrt{(1-\eta)p_3\lambda} - \lambda$ and $\mu_H \equiv \sqrt{(1+\eta)p_3\lambda} - \lambda$. Note that $\mu_H > \mu_L$. If $r \in (0,p_3)$, then $\partial \Pi_P/\partial \mu < 0$, therefore $((w,p_3),\mu_L) \succeq ((w,p_3),\mu_H)$. If the principal offers a contract (the conditions are discussed in Proposition 4.18 that follows), then by Definition 2.3 only μ_L leads to admissible solutions. Thus we obtain (a). If $r > p_3$, then $\partial \Pi_P/\partial \mu > 0$, therefore $((w,p_3),\mu_H) \succeq ((w,p_3),\mu_L)$. If the principal offers a contract (see Proposition 4.18), then only μ_H leads to admissible solutions. Therefore we obtain (c). If $r = p_3$, then $\partial \Pi_P/\partial \mu = 0$, indicating that the principal receives the same expected profit rate when the agent installs capacity μ_L or μ_H. If the principal offers a contract (see Proposition 4.18), then both μ_L and μ_H lead to admissible solutions and we obtain (b). □

Notation:

$$r_1 \equiv \eta p_3 + (1-\eta)\sqrt{p_2 p_3}\,,\ r_2 \equiv \left(1 + 2\eta\left(\frac{\sqrt{p_3}-\sqrt{p_2}}{\sqrt{p_2}}\right)\right) p_3,$$

$$\text{and } r_3 \equiv (1+2\eta)p_3 \tag{4.12}$$

Note that r_1, r_2 and r_3 are functions of λ and η. However we suppress the parameters (λ, η).

Define p_{cu} as follows[1]:

$$p_{cu} \equiv \frac{1}{9a^2}\left(b + C + \overline{C}\right)^2 \tag{4.13}$$

where $a \equiv 2\eta$, $b \equiv (1-2\eta)\sqrt{p_2}$, and $d \equiv -r\sqrt{p_2}$ and

$$C \equiv \sqrt[3]{\frac{\Delta_1 + \sqrt{\Delta_1^2 - 4\Delta_0^3}}{2}},\ \overline{C} \equiv \sqrt[3]{\frac{\Delta_1 - \sqrt{\Delta_1^2 - 4\Delta_0^3}}{2}},$$

$$\text{where } \Delta_0 \equiv b^2,\ \Delta_1 \equiv 2b^3 + 27a^2 d$$

Replacing Δ_0 and Δ_1 by the expressions of a, b and d we have

$$C = \sqrt[3]{\frac{2(1-2\eta)^3\sqrt{p_2^3} - 108\eta^2 r\sqrt{p_2} + \sqrt{-432\eta^2 r(1-2\eta)^3 p_2^2 + 11664\eta^4 r^2 p_2}}{2}}$$

and

$$\overline{C} = \sqrt[3]{\frac{2(1-2\eta)^3\sqrt{p_2^3} - 108\eta^2 r\sqrt{p_2} - \sqrt{-432\eta^2 r(1-2\eta)^3 p_2^2 + 11664\eta^4 r^2 p_2}}{2}}$$

We introduce several technical lemmas with proofs in the Appendix.

Lemma 4.13. *Let* $4/5 > \eta > 0$ *and* $\lambda > 0$.

(a) $\eta p_3 + (1-\eta)\sqrt{p_2 p_3} > p_2 > 0$.
(b) $p_3 > \eta p_3 + (1-\eta)\sqrt{p_2 p_3}$.
(c) $\left(1 + 2\eta\left(\sqrt{p_3} - \sqrt{p_2}\right)/\sqrt{p_2}\right) p_3 > p_3$.
(d) $(1+2\eta)p_3 > \left(1 + 2\eta\left(\sqrt{p_3} - \sqrt{p_2}\right)/\sqrt{p_2}\right) p_3$.

Lemma 4.13 implies that for $\eta \in (0, 4/5)$ we have $r_3 > r_2 > p_3 > r_1 > p_2$.

[1]The subscript "*cu*" stands for "cubic" because (4.13) is the square of the solution to Eq. (A.1), which is a cubic equation that is introduced later in the proof for Lemma 4.16.

Lemma 4.14. *Let $\eta > 0$ and $\lambda > 0$.*

(a) *If* $\left(1+3\eta+2\sqrt{\eta(1+2\eta)}\right)\lambda/(1+\eta) > r > \left(1+3\eta-2\sqrt{\eta(1+2\eta)}\right)\lambda/(1+\eta)$ *then* $0 > r-2\sqrt{(1+2\eta)r\lambda/(1+\eta)}+\lambda$.
(b) *If* $\left(1+3\eta-2\sqrt{\eta(1+2\eta)}\right)\lambda/(1+\eta) > r > 0$ *or* $r > \left(1+3\eta+2\sqrt{\eta(1+2\eta)}\right)\lambda/(1+\eta)$ *then* $r-2\sqrt{(1+2\eta)r\lambda/(1+\eta)}+\lambda > 0$.
(c) *If* $r = \left(1+3\eta-2\sqrt{\eta(1+2\eta)}\right)\lambda/(1+\eta)$ *or* $r = \left(1+3\eta+2\sqrt{\eta(1+2\eta)}\right)\lambda/(1+\eta)$ *then* $r-2\sqrt{(1+2\eta)r\lambda/(1+\eta)}+\lambda = 0$.

Lemma 4.15. *Given* $1 > \eta > 0$ *and* $\lambda > 0$, *then* $(1+2\eta)p_3 > \left(1+3\eta+2\sqrt{\eta(1+2\eta)}\right)\lambda/(1+\eta)$.

Lemma 4.16. *Consider* $\max_{x\in[\sqrt{p_2},\sqrt{p_3}]} f(x)$ *where* $f(x) = r+\lambda-\eta x^2-\sqrt{p_2}\left((1-2\eta)x+r/x\right)$ *and denote* $x^* = \operatorname{argmax}_{x\in[\sqrt{p_2},\sqrt{p_3}]} f(x)$. *The solutions to this optimization problem are*

(a) $x^* = \sqrt{p_2}$ *if* $r \in (0, p_2]$.
(b) $x^* = \sqrt{p_{cu}} \in \left(\sqrt{p_2}, \sqrt{p_3}\right)$ *if* $r \in (p_2, r_2)$.
(c) $x^* = \sqrt{p_3}$ *if* $r \geq r_2$.

Lemma 4.17. *Consider* $\max_{x\geq\sqrt{p_3}} f(x)$ *where* $f(x) = r+\lambda-\sqrt{p_1}\left((1+2\eta)x+r/x\right)$ *and denote* $x^* = \operatorname{argmax}_{x\geq\sqrt{p_3}} f(x)$. *The solutions to this optimization problem are*

(a) $x^* = \sqrt{p_3}$ *if* $r \in (0, r_3]$.
(b) $x^* = \sqrt{r/(1+2\eta)}$ *if* $r > r_3$.

Now we state Proposition 4.18, which serves as a stepping stone towards the main results Theorem 4.19 and Proposition 4.20 that follow later. Proposition 4.18 provides the optimal w^* and optimal p^* under some restrictions. These restrictions are later removed in the main results Theorem 4.19 and Proposition 4.20.

Recall that Proposition 4.9 describes the agent's optimal response to each contract offer (w, p). Since the principal will not propose a contract that is going to be rejected by a weakly risk-averse (WRA) agent, therefore Proposition 4.18 only considers pairs (w, p) that result in agent's non-negative expected utility rate. Define:

$$\begin{aligned}
\mathfrak{D}_{(4.7)} &\equiv \{(w,p) \text{ that satisfies (4.7) when } \eta \in (0,4/5)\}\\
\mathfrak{D}_{(4.8)} &\equiv \{(w,p) \text{ that satisfies (4.8) when } \eta \in (0,4/5)\}\\
\mathfrak{D}_{(4.9)} &\equiv \{(w,p) \text{ that satisfies (4.9) when } \eta \in (0,4/5)\}\\
\mathfrak{D}_{(4.10)} &\equiv \{(w,p) \text{ that satisfies (4.10) when } \eta \in (0,4/5)\}\\
\mathfrak{D}_{\text{WRA}} &\equiv \mathfrak{D}_{(4.7)} \cup \mathfrak{D}_{(4.8)} \cup \mathfrak{D}_{(4.9)} \cup \mathfrak{D}_{(4.10)}
\end{aligned} \tag{4.14}$$

Proposition 4.18. *Given a weakly risk-averse agent;*

(a) *If* $(w,p) \in \mathfrak{D}_{(4.7)}$, *then the principal does not propose a contract.*
(b) *Consider offers* $(w,p) \in \mathfrak{D}_{(4.8)} \cup \mathfrak{D}_{(4.9)}$.

(b1) *If* $r \in (0,p_2]$, *then the principal does not propose a contract.*
(b2) *If* $r \in (p_2,p_3]$, *then the principal offers* $(w^*,p^*) = \left(\eta p_{cu} + 2\sqrt{(1-\eta)p_{cu}\lambda} - \lambda, p_{cu}\right)$ *and the agent installs service capacity* $\mu^*(w^*,p^*) = \sqrt{(1-\eta)p_{cu}\lambda} - \lambda$.
(b3) *If* $r \in (p_3,r_2)$, *then the principal either offers* $(w^*,p^*) = (w_3,p_3)$ *and the agent installs* $\mu^*(w,p) = \sqrt{(1+\eta)p_3\lambda} - \lambda$, *or offers* $(w^*,p^*) = \left(\eta p_{cu} + 2\sqrt{(1-\eta)p_{cu}\lambda} - \lambda, p_{cu}\right)$ *and the agent installs service capacity* $\mu^*(w^*,p^*) = \sqrt{(1-\eta)p_{cu}\lambda} - \lambda$.
(b4) *If* $r \geq r_2$, *then the principal's offer is* $(w^*,p^*) = (w_3,p_3)$ *and the agent installs service capacity* $\mu^*(w^*,p^*) = \sqrt{(1+\eta)p_3\lambda} - \lambda$.

(c) *Consider offers* $(w,p) \in \mathfrak{D}_{(4.9)} \cup \mathfrak{D}_{(4.10)}$.

(c1) *If* $r \in (0,r_1]$, *then the principal does not propose a contract.*
(c2) *If* $r \in (r_1,p_3]$, *the principal offers a contract with* $(w^*,p^*) = (w_3,p_3)$ *and the agent installs service capacity* $\mu^*(w^*,p^*) = \sqrt{(1-\eta)p_3\lambda} - \lambda$.
(c3) *If* $r \in (p_3,r_3]$, *the principal offers a contract with* $(w^*,p^*) = (w_3,p_3)$ *and the agent installs service capacity* $\mu^*(w^*,p^*) = \sqrt{(1+\eta)p_3\lambda} - \lambda$.
(c4) *If* $r > r_3$, *the principal offers* $(w^*,p^*) = \left(2\sqrt{(1+\eta)r\lambda/(1+2\eta)} - \lambda, r/(1+2\eta)\right)$ *and the agent installs service capacity* $\mu^*(w^*,p^*) = \sqrt{(1+\eta)r\lambda/(1+2\eta)} - \lambda$.

Proof. The structure of the proof for Proposition 4.18 is depicted in Fig. 4.7.

Case $(w,p) \in \mathfrak{D}_{(4.7)}$: According to Proposition 4.9 part (a), in case the principal makes an offer, the agent accepts the contract but does not install any service capacity. Since $\partial\Pi_P/\partial w = -1 < 0$, thus $w^* = p$ and from Eq. (3.3) $\Pi_P(w^*,p;\mu^*(w^*,p)) = -w^* + p = -p + p = 0$. Therefore the principal does not propose a contract.

Case $(w,p) \in \mathfrak{D}_{(4.8)} \cup \mathfrak{D}_{(4.9)}$: According to Proposition 4.9 part (b), if $(w,p) \in \mathfrak{D}_{(4.8)}$, then in case the principal makes an offer, the agent accepts the contract and installs $\sqrt{(1-\eta)p\lambda} - \lambda$. Since $\partial\Pi_P/\partial w = -1 < 0$, therefore $w^* = \eta p + 2\sqrt{(1-\eta)p\lambda} - \lambda$. According to Propositions 4.9 part (c) and 4.12, if $(w,p) \in \mathfrak{D}_{(4.9)}$ (which implies $p = p_3$), then in case the principal makes an offer, the agent accepts the contract and installs $\sqrt{(1-\eta)p_3\lambda} - \lambda$ if $r \in (0,p_3)$, installs either $\sqrt{(1-\eta)p_3\lambda} - \lambda$ or $\sqrt{(1+\eta)p_3\lambda} - \lambda$ if $r = p_3$, or installs $\sqrt{(1+\eta)p_3\lambda} - \lambda$ if $r > p_3$. Since $\partial\Pi_P/\partial w = -1 < 0$, therefore $w^* = w_3$. Denote the principal's expected profit rate when $(w,p) = (w_3,p_3)$ and $\mu = \sqrt{(1-\eta)p_3\lambda} - \lambda$ by

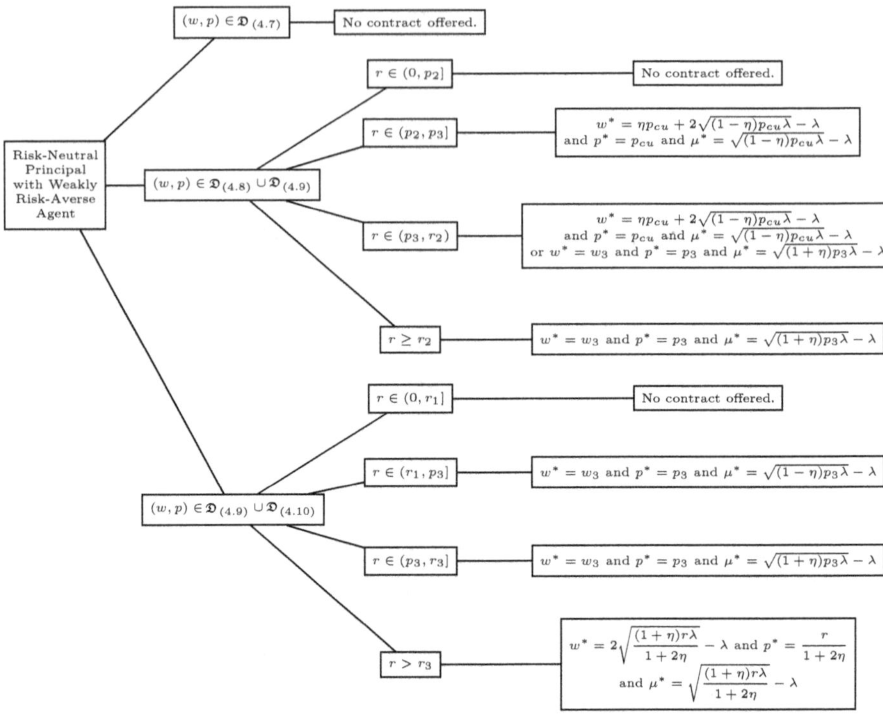

Fig. 4.7 Structure of the proof for Proposition 4.18

$\Pi_P^L(p_3)$, and denote the principal's expected profit rate when$(w, p) = (w_3, p_3)$ and $\mu = \sqrt{(1+\eta)p_3\lambda} - \lambda$ by $\Pi_P^H(p_3)$. By plugging the value of w, p and μ into Eq. (3.3):

$$\Pi_P^L(p_3) = r + \lambda - \eta p_3 - \sqrt{p_2}\left((1-2\eta)\sqrt{p_3} + \frac{r}{\sqrt{p_3}}\right)$$

$$= \left(\frac{\sqrt{p_3} - \sqrt{p_2}}{\sqrt{p_3}}\right)(r - r_1) \tag{4.15}$$

$$\Pi_P^H(p_3) = r + \lambda - \sqrt{p_1}\left((1+2\eta)\sqrt{p_3} + \frac{r}{\sqrt{p_3}}\right) \tag{4.16}$$

and the principal's optimization problem is $\max_{p \in [p_2, p_3]} \Pi_P(w^*, p; \mu^*(w^*, p))$ where:

$$\Pi_P(w^*, p; \mu^*(w^*, p)) = \begin{cases} r + \lambda - \eta p - \sqrt{p_2}\left((1-2\eta)\sqrt{p} + \frac{r}{\sqrt{p}}\right), \\ \quad \text{for } p \in [p_2, p_3) \\ \max\left\{\Pi_P^L(p_3), \Pi_P^H(p_3)\right\}, \text{ for } p = p_3 \end{cases}$$

Define $x \equiv \sqrt{p}$, the expression $r + \lambda - \eta p - \sqrt{p_2}\left((1-2\eta)\sqrt{p} + r/\sqrt{p}\right)$ can be restated as $f(x) = r + \lambda - \eta x^2 - \sqrt{p_2}\left((1-2\eta)x + r/x\right)$. Maximizing $f(x)$ with respect to x over $\left[\sqrt{p_2}, \sqrt{p_3}\right]$ is equivalent to maximizing $r + \lambda - \eta p - \sqrt{p_2}\left((1-2\eta)\sqrt{p} + r/\sqrt{p}\right)$ with respect to p over the interval $[p_2, p_3]$ in the sense that

$$\underset{p\in[p_2,p_3]}{\operatorname{argmax}}\left\{r + \lambda - \eta p - \sqrt{p_2}\left((1-2\eta)\sqrt{p} + \frac{r}{\sqrt{p}}\right)\right\} = \left(\underset{x\in\left[\sqrt{p_2},\sqrt{p_3}\right]}{\operatorname{argmax}} f(x)\right)^2$$

According to Lemma 4.13, $r_2 > p_3 > p_2$, therefore we examine the following subcases.

Subcase $r \in (0, p_2]$: According to Lemma 4.16 part (a), $p^* = p_2$; this case is taken care of in the case when $(w,p) \in \mathfrak{D}_{(4.7)}$ and the principal does not propose a contract.

Subcase $r \in (p_2, p_3]$: According to Lemma 4.16 part (b) and Proposition 4.12 part (a) and (b), $p^* = p_{cu}$ and the principal's expected profit rate is $\Pi_P(w^*, p^*; \mu^*(w^*, p^*)) > \Pi_P(p_2, p_2; 0) = 0$. Thus the principal proposes $w^* = \eta p_{cu} + 2\sqrt{(1-\eta)p_{cu}\lambda} - \lambda$ and $p^* = p_{cu}$ that induces the agent to install $\mu^*(w^*, p^*) = \sqrt{(1-\eta)p_{cu}\lambda} - \lambda$.

Subcase $r \in (p_3, r_2)$: According to Lemma 4.16 part (b) and Proposition 4.12 part (c), the principal chooses either $p^* = p_{cu}$ with expected profit rate $\Pi_P(w^*, p^*; \mu^*(w^*, p^*)) = r + \lambda - \eta p_{cu} - \sqrt{p_2}\left((1-2\eta)\sqrt{p_{cu}} + r/\sqrt{p_{cu}}\right) > \Pi_P(p_2, p_2; 0) = 0$, or chooses $p^* = p_3$ with expected profit rate $\Pi_P(w^*, p^*; \mu^*(w^*, p^*)) = \Pi_P^H(p_3) > \Pi_P^L(p_3) = \left(\sqrt{p_3} - \sqrt{p_2}\right)(r - r_1)/\sqrt{p_3} > 0$. However due to the difficulty of computing p_{cu} we do not explicitly identify the principal's optimal offer.

Subcase $r \geq r_2$: According to Lemma 4.16 part (c), $p^* = p_3$. According to Proposition 4.12 part (c) the agent installs capacity $\sqrt{(1+\eta)p_3\lambda} - \lambda$ and the principal's expected profit rate is $\Pi_P(w^*, p^*; \mu^*(w^*, p^*)) = \Pi_P^H(p_3) > \Pi_P^L(p_3) = \left(\sqrt{p_3} - \sqrt{p_2}\right)(r - r_1)/\sqrt{p_3} > 0$. Therefore the principal proposes $w^* = 2\sqrt{(1+\eta)p_3\lambda} - \lambda$ and $p^* = p_3$ that induces the agent to install $\mu^*(w^*, p^*) = \sqrt{(1+\eta)p_3\lambda} - \lambda$.

Case $(w,p) \in \mathfrak{D}_{(4.9)} \cup \mathfrak{D}_{(4.10)}$: According to Proposition 4.9 part (d), if $(w,p) \in \mathfrak{D}_{(4.10)}$, then in case the principal makes an offer, the agent accepts the contract and installs $\sqrt{(1+\eta)p\lambda} - \lambda$. Since $\partial\Pi_P/\partial w = -1 < 0$, therefore $w^* = 2\sqrt{(1+\eta)p\lambda} - \lambda$. According to Propositions 4.9 part (c) and 4.12, if $(w,p) \in \mathfrak{D}_{(4.9)}$ (which implies $p = p_3$), then in case the principal makes an offer, the agent accepts the contract and installs $\sqrt{(1-\eta)p_3\lambda} - \lambda$ if $r \in (0, p_3)$, installs either $\sqrt{(1-\eta)p_3\lambda} - \lambda$ or $\sqrt{(1+\eta)p_3\lambda} - \lambda$ if $r = p_3$, or installs $\sqrt{(1+\eta)p_3\lambda} - \lambda$ if $r > p_3$. Since $\partial\Pi_P/\partial w = -1 < 0$, therefore $w^* = w_3$. Recall the definition of $\Pi_P^L(p_3)$ and $\Pi_P^H(p_3)$ (see Eqs. (4.15) and (4.16)). The principal's optimization problem is $\max_{p\geq p_3} \Pi_P(w^*, p; \mu^*(w^*, p))$ where:

$$\Pi_P(w^*, p; \mu^*(w^*, p)) = \begin{cases} \max\left\{\Pi_P^L(p_3), \Pi_P^H(p_3)\right\}, \text{ for } p = p_3 \\ r + \lambda - \sqrt{p_1}\left((1+2\eta)\sqrt{p} + \frac{r}{\sqrt{p}}\right), \text{ for } p > p_3 \end{cases}$$

Define $x \equiv \sqrt{p}$, the expression $r + \lambda - \sqrt{p_1}\left((1+2\eta)\sqrt{p} + r/\sqrt{p}\right)$ can be restated as $f(x) = r + \lambda - \sqrt{p_1}\left((1+2\eta)x + r/x\right)$. Maximizing $f(x)$ for $x \geq \sqrt{p_3}$ is equivalent to maximizing $r + \lambda - \sqrt{p_1}\left((1+2\eta)\sqrt{p} + r/\sqrt{p}\right)$ for $p \geq p_3$ in the sense that

$$\operatorname*{argmax}_{p\geq p_3}\left\{r + \lambda - \sqrt{p_1}\left((1+2\eta)\sqrt{p} + \frac{r}{\sqrt{p}}\right)\right\} = \left(\operatorname*{argmax}_{x\geq\sqrt{p_3}} f(x)\right)^2$$

According to Lemma 4.13, $r_3 > p_3 > r_1$, therefore we examine the following subcases.

Subcase $r \in (0, r_1]$: According to Lemma 4.17 part (a), $p^* = p_3$. By Proposition 4.12 part (a), $\Pi_P(w^*, p^*; \mu^*(w^*, p^*)) = \Pi_P^L(p_3) = \left(\sqrt{p_3} - \sqrt{p_2}\right)(r - r_1)/\sqrt{p_3}$ and note that $\Pi_P(w^*, p^*; \mu^*(w^*, p^*)) \leq 0$, therefore the principal does not propose a contract.

Subcase $r \in (r_1, p_3]$: According to Lemma 4.17 part (a), $p^* = p_3$. According to Proposition 4.12 part (a) and (b), the principal's expected profit rate is $\Pi_P(w^*, p^*; \mu^*(w^*, p^*)) = \Pi_P^L(p_3) = \left(\sqrt{p_3} - \sqrt{p_2}\right)(r - r_1)/\sqrt{p_3} > 0$, therefore the principal proposes a contract with $w^* = w_3$ and $p^* = p_3$ that induces the agent to install $\mu^*(w^*, p^*) = \sqrt{(1-\eta)p_3\lambda} - \lambda$.

Subcase $r \in (p_3, r_3]$: According to Lemma 4.17 part (a), $p^* = p_3$. According to Proposition 4.12 part (c), the principal's expected profit rate is $\Pi_P(w^*, p^*; \mu^*(w^*, p^*)) = \Pi_P^H(p_3) > \Pi_P^L(p_3) = \left(\sqrt{p_3} - \sqrt{p_2}\right)(r - r_1)/\sqrt{p_3} > 0$, therefore the principal proposes a contract with $w^* = w_3$ and $p^* = p_3$ that induces the agent to install $\mu^*(w^*, p^*) = \sqrt{(1+\eta)p_3\lambda} - \lambda$.

Subcase $r > r_3$: According to Lemma 4.17 part (b), $p^* = r/(1+2\eta)$ and the principal's expected profit rate is $\Pi_P(w^*, p^*; \mu^*(w^*, p^*)) = r - 2\sqrt{(1+2\eta)r\lambda/(1+\eta)} + \lambda$. According to Lemmas 4.14 and 4.15, $\Pi_P(w^*, p^*; \mu^*(w^*, p^*)) > 0$, therefore the principal proposes a contract with $w^* = 2\sqrt{(1+\eta)r\lambda/(1+2\eta)} - \lambda$ and $p^* = r/(1+2\eta)$ that induces the agent to install service capacity $\mu^*(w^*, p^*) = \sqrt{(1+\eta)r\lambda/(1+2\eta)} - \lambda$.

□

We describe the principal's optimal strategy in Theorem 4.19 and Proposition 4.20. We identify the principal's optimal offer only when $r \in (0, p_3]$ or $r \geq r_2$, (see Theorem 4.19). The cases when $r \in (p_3, r_2)$ are discussed in Proposition 4.20. We prove Theorem 4.19 and Proposition 4.20 together.

Theorem 4.19. *Consider a weakly risk-averse agent and* $(w, p) \in \mathfrak{D}_{\text{WRA}}$.

(a) *If $r \in (0, p_2]$, then the principal does not propose a contract.*
(b) *If $r \in (p_2, p_3]$, then the principal's offer and the capacity installed by the agent are*

$$(w^*, p^*) = \left(\eta p_{cu} + 2\sqrt{(1-\eta)p_{cu}\lambda} - \lambda, p_{cu}\right) \text{ and}$$
$$\mu^*(w^*, p^*) = \sqrt{(1-\eta)p_{cu}\lambda} - \lambda \tag{4.17}$$

and the principal's expected profit rate is

$$\Pi_P(w^*, p^*; \mu^*(w^*, p^*)) = r + \lambda - \eta p_{cu} - \sqrt{p_2}\left((1-2\eta)\sqrt{p_{cu}} + \frac{r}{\sqrt{p_{cu}}}\right) \tag{4.18}$$

(c) *If $r \in [r_2, r_3]$, then the principal's offer and the capacity installed by the agent are*

$$(w^*, p^*) = (w_3, p_3) \text{ and } \mu^*(w^*, p^*) = \sqrt{(1+\eta)p_3\lambda} - \lambda \tag{4.19}$$

and the principal's expected profit rate is

$$\Pi_P(w^*, p^*; \mu^*(w^*, p^*)) = r + \lambda - \sqrt{p_1}\left((1+2\eta)\sqrt{p_3} + \frac{r}{\sqrt{p_3}}\right) \tag{4.20}$$

(d) *If $r > r_3$, then the principal's offer and the capacity installed by the agent are*

$$(w^*, p^*) = \left(2\sqrt{\frac{(1+\eta)r\lambda}{1+2\eta}} - \lambda, \frac{r}{1+2\eta}\right) \text{ and}$$
$$\mu^*(w^*, p^*) = \sqrt{\frac{(1+\eta)r\lambda}{1+2\eta}} - \lambda \tag{4.21}$$

and the principal's expected profit rate is

$$\Pi_P(w^*, p^*; \mu^*(w^*, p^*)) = r - 2\sqrt{\frac{(1+2\eta)r\lambda}{1+\eta}} + \lambda \tag{4.22}$$

Proposition 4.20. *Given a weakly risk-averse agent and $(w, p) \in \mathfrak{D}_{\text{WRA}}$. If $r \in (p_3, r_2)$, then either*

$$(w^*, p^*) = \left(\eta p_{cu} + 2\sqrt{(1-\eta)p_{cu}\lambda} - \lambda, p_{cu}\right) \text{ and}$$
$$\mu^*(w^*, p^*) = \sqrt{(1-\eta)p_{cu}\lambda} - \lambda$$

resulting in principal's expected profit rate

$$\Pi_P(w^*,p^*;\mu^*(w^*,p^*)) = r + \lambda - \eta p_{cu} - \sqrt{p_2}\left((1-2\eta)\sqrt{p_{cu}} + r/\sqrt{p_{cu}}\right)$$

or the principal offers and the agent installs

$$(w^*,p^*) = (w_3,p_3) \text{ and } \mu^*(w^*,p^*) = \sqrt{(1+\eta)p_3\lambda} - \lambda$$

resulting in principal's expected utility rate

$$\Pi_P(w^*,p^*;\mu^*(w^*,p^*)) = r + \lambda - \sqrt{p_1}\left((1+2\eta)\sqrt{p_3} + r/\sqrt{p_3}\right)$$

Proof. In part (b) of Proposition 4.18, we solved for (w^*,p^*) by restricting r to be in $(0,p_2]$, or in $(p_2,p_3]$, or in (p_3,r_2), or in $[r_2,+\infty)$. In part (c) of Proposition 4.18, we solved for (w^*,p^*) by restricting r to be in $(0,r_1]$, or in $(r_1,p_3]$, or in $(p_3,r_3]$, or in $(r_3,+\infty)$. The principal maximizes her expected profit rate by offering contract that lead to admissible solutions (Definition 2.3) for any given value of r, η and λ. The structure of the proof for Theorem 4.19 and Proposition 4.20 is depicted in Fig. 4.8.

Case $r \in (0,p_2]$: According to Proposition 4.18 part (a), (b1) and (c1), the principal does not propose a contract. This case corresponds to Theorem 4.19 (a).

Case $r \in (p_2,p_3]$: If $r \in (p_2,r_1]$, then according to Proposition 4.18 part (a), (b2) and (c1), the principal offers $(w^*,p^*) = \left(\eta p_{cu} + 2\sqrt{(1-\eta)p_{cu}\lambda} - \lambda, p_{cu}\right)$

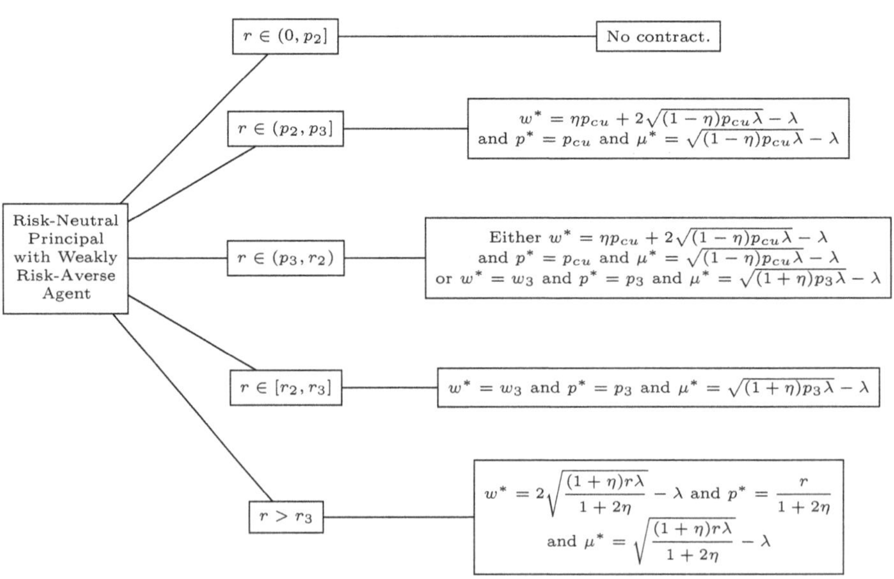

Fig. 4.8 Structure of the proof for Theorem 4.19 and Proposition 4.20

and the agent installs $\mu^*(w^*,p^*) = \sqrt{(1-\eta)p_{cu}\lambda} - \lambda$. If $r \in (r_1, p_3]$, then according to Proposition 4.18 part (a), (b2) and (c2) and Lemma 4.16 part (b) the principal offers $(w^*,p^*) = \left(\eta p_{cu} + 2\sqrt{(1-\eta)p_{cu}\lambda} - \lambda, p_{cu}\right)$ and the agent installs $\mu^*(w^*,p^*) = \sqrt{(1-\eta)p_{cu}\lambda} - \lambda$. This case is addressed in Theorem 4.19 (b).

Case $r \in (p_3, r_2)$: According to Proposition 4.18 part (a), (b3) and (c3), the principal either offers a contract $(w^*,p^*) = \left(\eta p_{cu} + 2\sqrt{(1-\eta)p_{cu}\lambda} - \lambda, p_{cu}\right)$ and the agent installs $\mu^*(w^*,p^*) = \sqrt{(1-\eta)p\lambda} - \lambda$, or offers $(w^*,p^*) = (w_3, p_3)$ and the agent installs $\mu^*(w^*,p^*) = \sqrt{(1+\eta)p_3\lambda} - \lambda$. This case corresponds to Proposition 4.20.

Case $r \in [r_2, r_3]$: According to Proposition 4.18 part (a), (b4) and (c3), the principal offers a contract with $(w^*,p^*) = (w_3, p_3)$ and the agent installs $\mu^*(w^*,p^*) = \sqrt{(1+\eta)p_3\lambda} - \lambda$. This case corresponds to Theorem 4.19 (c).

Case $r > r_3$: According to Proposition 4.18 part (a), (b4) and (c4) and Lemma 4.17 part (b), the principal offers a contract with $(w^*,p^*) = \left(2\sqrt{(1+\eta)r\lambda/(1+2\eta)} - \lambda, r/(1+2\eta)\right)$ and the agent installs service capacity $\mu^*(w^*,p^*) = \sqrt{(1+\eta)r\lambda/(1+2\eta)} - \lambda$. This case corresponds to Theorem 4.19 (d).

□

Theorem 4.19 and Proposition 4.20 indicate that the existence of a beneficial contract with a weakly risk-averse agent is determined exogenously by the revenue rate r, the failure rate λ, and the risk coefficient η.

Since it is difficult to identify the principal's optimal offer when $r \in (p_3, r_2)$ due to the difficulty of computing p_{cu} we resort to numerical results to better understand the principal's choices.

Remark 4.21. Figure 4.9 demonstrates that when $\eta = 0.1$ and $\eta = 0.5$ there exists an $r_0 \in (p_3, r_2)$ such that when $r \in (p_3, r_0)$, the principal offers $(w^*,p^*) = \left(\eta p_{cu} + 2\sqrt{(1-\eta)p_{cu}\lambda} - \lambda, p_{cu}\right)$, when $r \in (r_0, r_2)$, she offers $(w^*,p^*) = (w_3, p_3)$ and when $r = r_0$, the principal is indifferent about the two alternative offers. However due to the difficulty of computing p_{cu} (Eq. (4.13)), it is not clear how to determine the general existence of such an r_0 for all $\eta \in (0, 4/5)$ and identify an explicit expression of r_0 as a function of λ and η.

4.2 Optimal Strategies Given a Strongly Risk-Averse Agent

For the strongly risk-averse (SRA) agent we first derive the agent's optimal strategy. The agent's optimization problem is stated in (4.4).

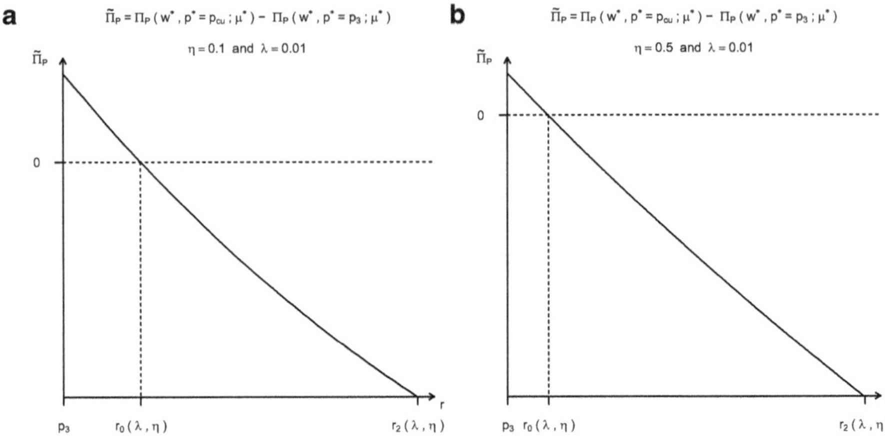

$\eta = 0.1$ $\eta = 0.5$

Fig. 4.9 The value of $\tilde{\Pi}_P \equiv \Pi_P(w^*, p^* = p_{cu}; \mu^*) - \Pi_P(w^*, p^* = p_3; \mu^*)$ for $r \in (p_3, r_2)$

Notation:

$$w_4 = p_4 \equiv \left(1 + 2\eta + 2\sqrt{\eta(1+\eta)}\right)\lambda \tag{4.23}$$

Note that w_4 and p_4 are functions of λ and η which are suppressed in our notation.

A technical lemma used later is introduced next (see proof in the Appendix).

Lemma 4.22. *Let $\eta > 1/3$ and $\lambda > 0$, then $\left(1 + 2\eta + 2\sqrt{\eta(1+\eta)}\right)\lambda > 4\lambda/(1+\eta)$.*

We describe a strongly risk-averse agent's optimal response to any possible offered contract $(w, p) \in \mathbb{R}^2_+$ in Proposition 4.23.

Proposition 4.23. *Consider a strongly risk-averse agent ($\eta \geq 4/5$).*

(a) *Given*

$$w \geq p \in (0, p_4) \tag{4.24}$$

then the agent would accept the contract if offered and install $\mu^*(w,p) = 0$ with resulting expected utility rate $u_A(\mu^*(w,p); w, p) = w - p \geq 0$. The agent rejects the contract if $p \in (0, p_4)$ and $w \in (0, p)$.

(b) *Given*

$$p = p_4 \text{ and } w \geq w_4 \tag{4.25}$$

then the agent would accept the contract if offered and is indifferent about installing either $\mu^*(w,p) = 0$ *or* $\mu^*(w,p) = \sqrt{(1+\eta)p_4\lambda} - \lambda$. *In both cases the agent's expected utility rate is* $u_A(\mu^*(w,p); w,p) = w - p_4 \geq 0$. *If* $r \in (0, p_4]$, *then neither* $\mu^* = 0$ *nor* $\mu^* = \sqrt{(1+\eta)p_4\lambda} - \lambda$ *leads to admissible solutions (see Definition 2.3). If* $r > p_4$, *then there exists* w^* *such that* $\left((w^*, p_4), \mu^* = \sqrt{(1+\eta)p_4\lambda} - \lambda\right)$ *is the unique admissible solution (for proof see Proposition 4.24). The agent rejects the contract if* $p = p_4$ *and* $w \in (0, w_4)$.

(c) *Given*

$$p > p_4 \text{ and } w \geq 2\sqrt{(1+\eta)p\lambda} - \lambda \tag{4.26}$$

then the agent would accept the contract if offered and install $\mu^*(w,p) = \sqrt{(1+\eta)p\lambda} - \lambda$ *with resulting expected utility rate* $u_A(\mu^*(w,p); w,p) = w - 2\sqrt{(1+\eta)p\lambda} + \lambda \geq 0$. *The agent rejects the contract if* $p > p_4$ *and* $w \in \left(0, 2\sqrt{(1+\eta)p\lambda} - \lambda\right)$.

Proof. According to Table 4.1, the optimization of $u(\mu)$ when $\eta \in [4/5, 1)$ versus $\eta \geq 1$ is different. Therefore we prove the proposition separately for $\eta \in [4/5, 1)$ and $\eta \geq 1$.

Case $\eta \in [4/5, 1)$: Recall the definition of p_1 and p_2 in (4.5). Note that $4p_2 > p_2 > 4p_1$ and according to Lemmas 4.7 part (b) and (c) and 4.22, $p_2 \geq p_4 > 4p_1$. Therefore we have $4p_2 > p_2 \geq p_4 > 4p_1$. Figure 4.10 depicts the shape of $u(\mu)$ when $\eta \in [4/5, 1)$ and the value of p falls in different ranges. The structure of the proof when $\eta \in [4/5, 1)$ is depicted in Fig. 4.11.

Case $p \in (0, 4p_1]$: According to Table 4.1, $u(\mu)$ is decreasing for $\mu \geq 0$. Thus the agent's optimal service capacity is $\mu^*(w,p) = 0$ and from (4.3) $u(\mu^*(w,p)) = w - p$.

Subcase $w \in (0, p)$: $u(\mu^*(w,p)) < 0$, therefore the agent rejects the contract.
Subcase $w \geq p$: $u(\mu^*(w,p)) \geq 0$, thus the agent would accept the contract if offered.

Case $p \in (4p_1, p_2]$: According to Table 4.1, there is a service capacity that maximizes $u(\mu)$ for $\mu \in [0, \lambda)$ and a service capacity that maximizes $u(\mu)$ for $\mu > \lambda$. Denote the optimal service capacity in $[0, \lambda)$ by $\mu^*_{[0,\lambda)}(w,p)$. Note that $u(\mu)$ is decreasing with respect to μ over $[0, \lambda)$, therefore the agent's optimal service capacity is $\mu^*_{[0,\lambda)}(w,p) = 0$ and from (4.3) $u\left(\mu^*_{[0,\lambda)}(w,p)\right) = w - p$. Denote the optimal service capacity for $\mu > \lambda$ by $\mu^*_{\lambda}(w,p)$. From the first order condition $\mu^*_{\lambda}(w,p) = \sqrt{(1+\eta)p\lambda} - \lambda$ and from (4.3) $u\left(\mu^*_{\lambda}(w,p)\right) = w - 2\sqrt{(1+\eta)p\lambda} + \lambda$. The agent has a choice of two service capacities and he installs the one that generates a higher expected utility rate. Note that $u\left(\mu^*_{\lambda}(w,p)\right) - u\left(\mu^*_{[0,\lambda)}(w,p)\right) = p - 2\sqrt{(1+\eta)p\lambda} + \lambda$. According to

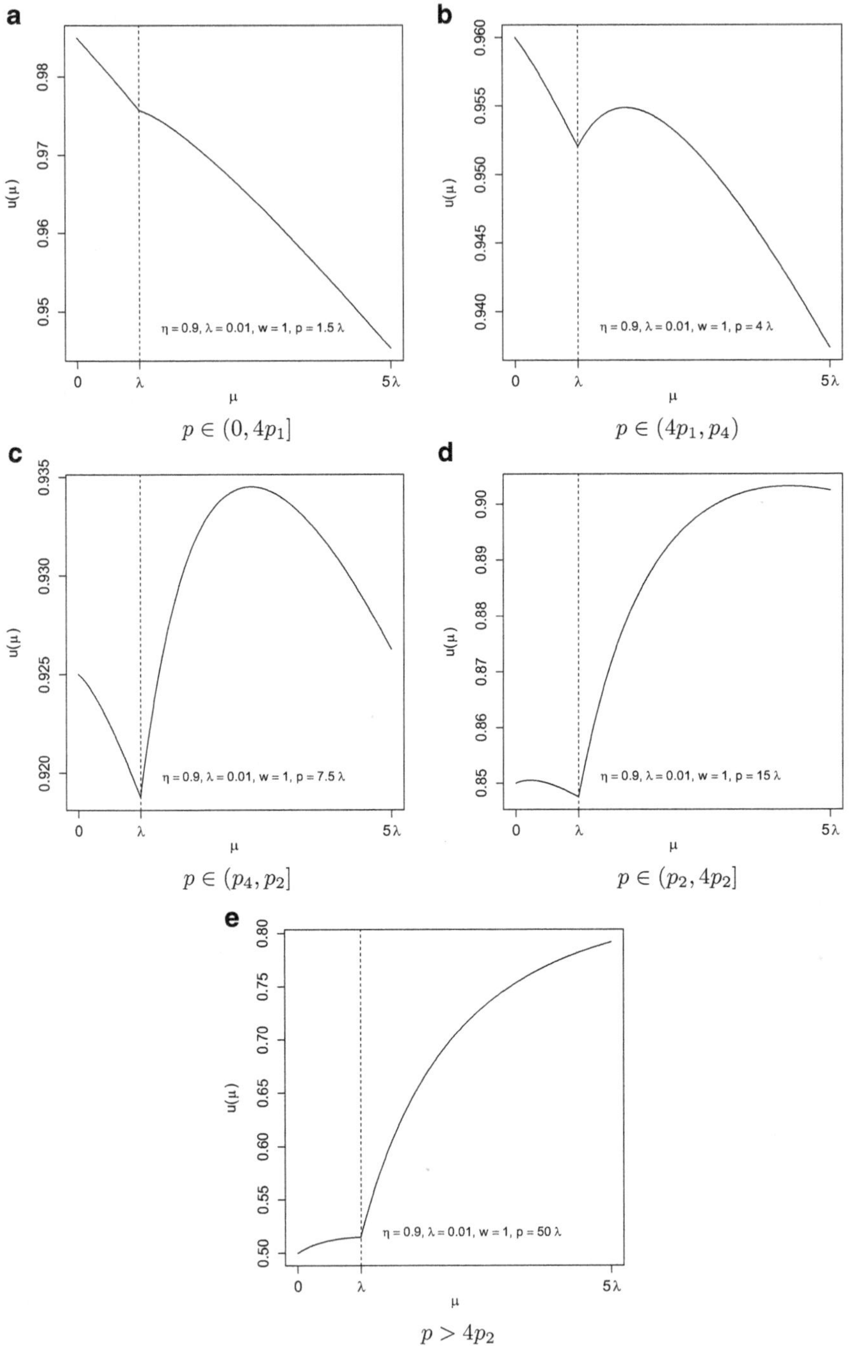

Fig. 4.10 Illustration of the forms of $u(\mu)$ when $\eta \in [4/5, 1)$

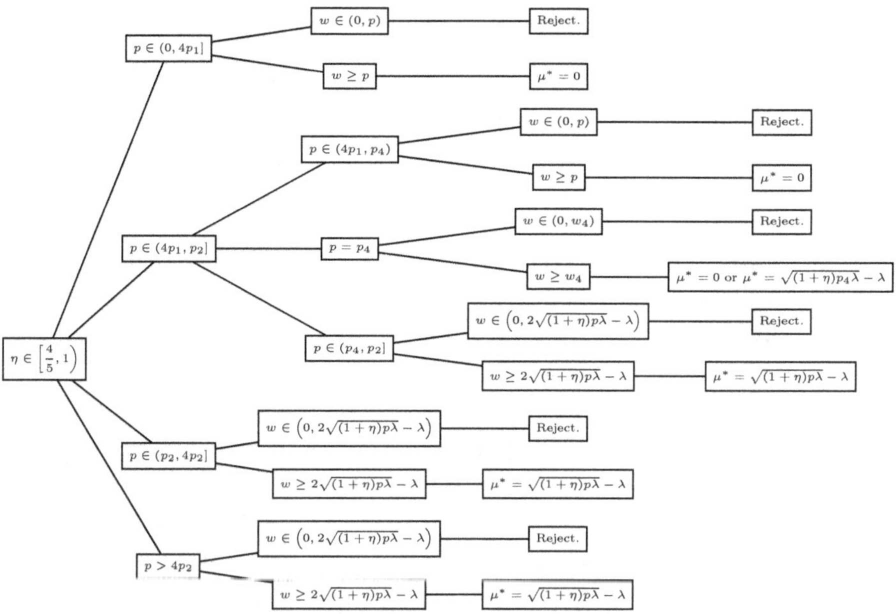

Fig. 4.11 Structure of the proof for Proposition 4.23 when $\eta \in [4/5, 1)$

Lemmas 4.7 part (b) and (c) and 4.22, $p_2 \geq p_4 > 4p_1$, therefore we examine the following subcases.

Subcase $p \in (4p_1, p_4)$: According to Lemma 4.6, $4p_1 > \left(1 + 2\eta - 2\sqrt{\eta(1+\eta)}\right)\lambda$ and according to Lemma 4.5 part (a), $u\left(\mu^*_{[0,\lambda)}(w,p)\right) > u\left(\mu^*_{\lambda}(w,p)\right)$, thus the agent's optimal service capacity is $\mu^*(w,p) = 0$ and $u(\mu^*(w,p)) = w - p$.

Subsubcase $w \in (0,p)$: $u(\mu^*(w,p)) < 0$, therefore the agent rejects the contract.

Subsubcase $w \geq p$: $u(\mu^*(w,p)) \geq 0$, therefore the agent would accept the contract if offered.

Subcase $p = p_4$: According to Lemma 4.5 part (c), $u\left(\mu^*_{[0,\lambda)}(w,p)\right) = u\left(\mu^*_{\lambda}(w,p)\right)$, indicating that installing $\mu^*_{[0,\lambda)}(w,p)$ or $\mu^*_{\lambda}(w,p)$ results in the same agent's expected utility rate. Therefore the agent is indifferent about installing $\mu^*(w,p) = 0$ or $\mu^*(w,p) = \sqrt{(1+\eta)p_4\lambda} - \lambda$ with expected utility rate $u(\mu^*(w,p); w,p) = w - w_4$. However the principal would not propose a contract in this case because none of these capacities leads to admissible solutions (see Definition 2.3). For proof see Proposition 4.24. According to Lemma 4.4, $p_4 > 4p_1 \Rightarrow w_4 = 2\sqrt{(1+\eta)p_4\lambda} - \lambda > 0$.

Subsubcase $w \in (0, w_4)$: $u(\mu^*(w,p)) < 0$, therefore the agent rejects the contract.

Subsubcase $w \geq w_4$: $u(\mu^*(w,p)) \geq 0$, therefore the agent would accept the contract if offered.

Subcase $p \in (p_4, p_2]$: From Lemma 4.5 part (b), $u\left(\mu_\lambda^*(w,p)\right) > u\big(\mu_{[0,\lambda)}^*(w,p)\big)$, thus the agent's optimal service capacity is $\mu^*(w,p) = \sqrt{(1+\eta)p\lambda} - \lambda$ and $u(\mu^*(w,p)) = w - 2\sqrt{(1+\eta)p\lambda} + \lambda$. According to Lemma 4.4, $p > p_4 > 4p_1 \Rightarrow 2\sqrt{(1+\eta)p\lambda} - \lambda > 0$, therefore we further examine the following subcases.

Subsubcase $w \in \left(0, 2\sqrt{(1+\eta)p\lambda} - \lambda\right)$: $u(\mu^*(w,p)) < 0$, therefore the agent rejects the contract.

Subsubcase $w \geq 2\sqrt{(1+\eta)p\lambda} - \lambda$: $u(\mu^*(w,p)) \geq 0$, therefore the agent would accept the contract if offered.

Case $p \in (p_2, 4p_2]$: According to Table 4.1, there is a service capacity that maximizes $u(\mu)$ for $\mu \in (0, \lambda]$ and a service capacity that maximizes $u(\mu)$ for $\mu > \lambda$. Denote the optimal service capacity in $(0,\lambda]$ by $\mu_{(0,\lambda]}^*(w,p)$. From the first order condition $\mu_{(0,\lambda]}^*(w,p) = \sqrt{(1-\eta)p\lambda} - \lambda$ and from (4.3) $u\left(\mu_{(0,\lambda]}^*(w,p)\right) = w - \eta p - 2\sqrt{(1-\eta)p\lambda} + \lambda$. Denote the optimal service capacity for $\mu > \lambda$ by $\mu_\lambda^*(w,p)$. From the first order condition $\mu_\lambda^*(w,p) = \sqrt{(1+\eta)p\lambda} - \lambda$ and from (4.3) $u\left(\mu_\lambda^*(w,p)\right) = w - 2\sqrt{(1+\eta)p\lambda} + \lambda$. The agent has to decide which of the two service capacities he installs and he chooses the one with higher expected utility rate. Note that $u\left(\mu_\lambda^*(w,p)\right) - u\left(\mu_{(0,\lambda]}^*(w,p)\right) = \eta p - 2\left(\sqrt{1+\eta} - \sqrt{1-\eta}\right)\sqrt{p\lambda}$. According to Lemma 4.8 part (b) and (c), $p > p_2 \geq p_3$, and according to Lemma 4.2 part (a) $u\left(\mu_\lambda^*(w,p)\right) > u\left(\mu_{(0,\lambda]}^*(w,p)\right)$, therefore the agent's optimal service capacity is $\mu^*(w,p) = \sqrt{(1+\eta)p\lambda} - \lambda$ and $u(\mu^*(w,p)) = w - 2\sqrt{(1+\eta)p\lambda} + \lambda$. According to Lemma 4.4, $p > p_2 > 4p_1 \Rightarrow 2\sqrt{(1+\eta)p\lambda} - \lambda > 0$, therefore we examine the following subcases.

Subcase $w \in \left(0, 2\sqrt{(1+\eta)p\lambda} - \lambda\right)$: $u(\mu^*(w,p)) < 0$, thus the agent rejects the contract.

Subcase $w \geq 2\sqrt{(1+\eta)p\lambda} - \lambda$: $u(\mu^*(w,p)) \geq 0$, therefore the agent would accept the contract if offered.

Case $p > 4p_2$: According to Table 4.1, the service capacity that maximizes $u(\mu)$ must satisfy $\mu > \lambda$. From the first order condition $\mu^*(w,p) = \sqrt{(1+\eta)p\lambda} - \lambda$ and $u(\mu^*(w,p)) = w - 2\sqrt{(1+\eta)p\lambda} + \lambda$. According to Lemma 4.4, $p > 4p_2 > 4p_1 \Rightarrow 2\sqrt{(1+\eta)p\lambda} - \lambda > 0$, therefore we examine the following subcases.

Subcase $w \in \left(0, 2\sqrt{(1+\eta)p\lambda} - \lambda\right)$: $u(\mu^*(w,p)) < 0$, thus the agent rejects the contract.

Subcase $w \geq 2\sqrt{(1+\eta)p\lambda} - \lambda$: $u(\mu^*(w,p)) \geq 0$, therefore the agent would accept the contract if offered.

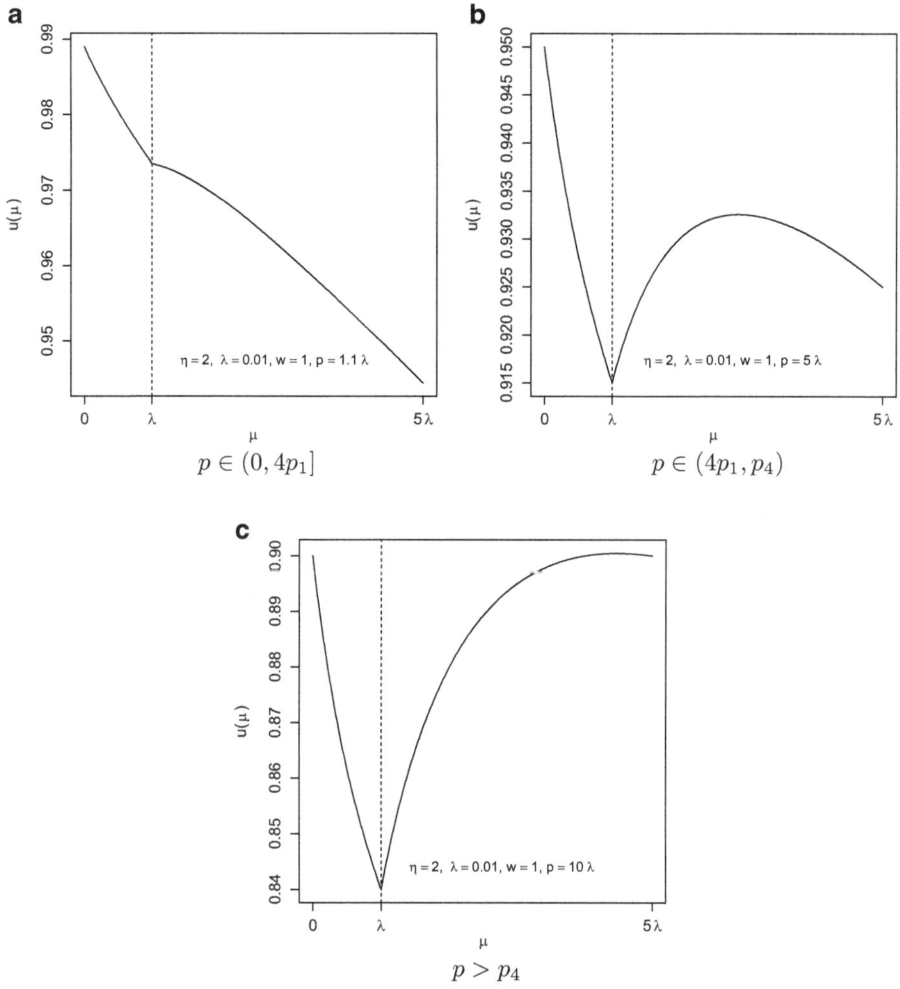

Fig. 4.12 Illustration of the forms of $u(\mu)$ when $\eta \geq 1$

This completes the proof for Proposition 4.23 when $\eta \in [4/5, 1)$.

Case $\eta \geq 1$: According to Lemma 4.22, $p_4 > 4p_1$. Figure 4.12 depicts the shape of $u(\mu)$ when $\eta \geq 1$ and the value of p falls in different ranges. The proof when $\eta \geq 1$ is depicted in Fig. 4.13.

Case $p \in (0, 4p_1]$: According to Table 4.1, $u(\mu)$ is decreasing with respect to $\mu \geq 0$. Thus the agent's optimal service capacity is $\mu^*(w, p) = 0$ and from (4.3) $u(\mu^*(w, p)) = w - p$.

Subcase $w \in (0, p)$: $u(\mu^*(w, p)) < 0$, therefore the agent rejects the contract.

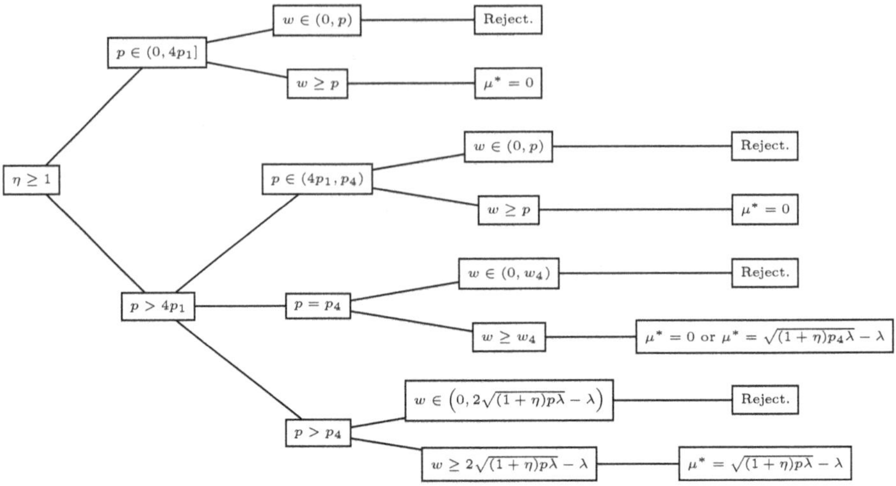

Fig. 4.13 Structure of the proof for Proposition 4.23 when $\eta \geq 1$

Subcase $w \geq p$: $u(\mu^*(w,p)) \geq 0$, thus the agent would accept the contract if offered.

Case $p > 4p_1$: According to Table 4.1, there is a service capacity that maximizes $u(\mu)$ for $\mu \in [0, \lambda)$ and a service capacity that maximizes $u(\mu)$ for $\mu > \lambda$. Denote the optimal service capacity in $[0, \lambda)$ by $\mu^*_{[0,\lambda)}(w,p)$. Note that $u(\mu)$ is decreasing with respect to μ over $[0, \lambda)$, therefore the agent's optimal service capacity is $\mu^*_{[0,\lambda)}(w,p) = 0$ and from (4.3) $u\left(\mu^*_{[0,\lambda)}(w,p)\right) = w - p$. Denote the optimal service capacity for $\mu > \lambda$ by $\mu^*_\lambda(w,p)$. From the first order condition $\mu^*_\lambda(w,p) = \sqrt{(1+\eta)p\lambda} - \lambda$ and from (4.3) $u\left(\mu^*_\lambda(w,p)\right) = w - 2\sqrt{(1+\eta)p\lambda} + \lambda$. The agent has to decide which of the two service capacities he is going to install and he chooses the one that generates a higher expected utility rate. Note that $u\left(\mu^*_\lambda(w,p)\right) - u\left(\mu^*_{[0,\lambda)}(w,p)\right) = p - 2\sqrt{(1+\eta)p\lambda} + \lambda$. According to Lemma 4.22, $p_4 > 4p_1$ and we need to examine the following subcases.

Subcase $p \in (4p_1, p_4)$: According to Lemma 4.6, $4p_1 > \left(1 + 2\eta - 2\sqrt{\eta(1+\eta)}\right)\lambda$ and according to Lemma 4.5 part (a), $u\left(\mu^*_{[0,\lambda)}(w,p)\right) > u\left(\mu^*_\lambda(w,p)\right)$, therefore the agent's optimal service capacity is $\mu^*(w,p) = 0$ and $u(\mu^*(w,p)) = w - p$.

Subsubcase $w \in (0,p)$: $u(\mu^*(w,p)) < 0$, therefore the agent rejects the contract.

Subsubcase $w \geq p$: $u(\mu^*(w,p)) \geq 0$, therefore the agent would accept the contract if offered.

Subcase $p = p_4$: According to Lemma 4.5 part (c), $u\left(\mu^*_{[0,\lambda)}(w,p)\right) = u\left(\mu^*_\lambda(w,p)\right)$, indicating that installing $\mu^*_{[0,\lambda)}(w,p)$ or $\mu^*_\lambda(w,p)$ leads to the same agent's expected utility rate. Therefore the agent is indifferent about installing $\mu^*(w,p) = 0$ or $\mu^*(w,p) = \sqrt{(1+\eta)p_4\lambda} - \lambda$ and in such case $u\left(\mu^*(w,p); w,p\right) = w - w_4$. However the principal would not propose a contract in this case, because none of these capacities leads to admissible solutions (see Definition 2.3). For proof see Proposition 4.24. According to Lemma 4.4, $p_4 > 4p_1 \Rightarrow w_4 = 2\sqrt{(1+\eta)p_4\lambda} - \lambda > 0$.

Subsubcase $w \in (0, w_4)$: $u(\mu^*(w,p)) < 0$, therefore the agent rejects the contract.

Subsubcase $w \geq w_4$: $u(\mu^*(w,p)) \geq 0$, therefore the agent would accept the contract if offered.

Subcase $p > p_4$: From Lemma 4.5 part (b), $u\left(\mu^*_\lambda(w,p)\right) > u\left(\mu^*_{[0,\lambda)}(w,p)\right)$, thus the agent's optimal capacity is $\mu^*(w,p) = \sqrt{(1+\eta)p\lambda} - \lambda$ and $u(\mu^*(w,p)) = w - 2\sqrt{(1+\eta)p\lambda} + \lambda$. According to Lemma 4.4, $p > p_4 > 4p_1 \Rightarrow 2\sqrt{(1+\eta)p\lambda} - \lambda > 0$, therefore we further examine the following subcases.

Subsubcase $w \in \left(0, 2\sqrt{(1+\eta)p\lambda} - \lambda\right)$: $u(\mu^*(w,p)) < 0$, thus the agent rejects the contract.

Subsubcase $w \geq 2\sqrt{(1+\eta)p\lambda} - \lambda$: $u(\mu^*(w,p)) \geq 0$, therefore the agent would accept the contract if offered.

□

In summary, given exogenous market conditions such that a mutually beneficial contract with a strongly risk-averse agent exists (see Theorem 4.27 later), only one formula is needed for the agent to compute his optimal service capacity: $\mu^*(w,p) = \sqrt{(1+\eta)p\lambda} - \lambda > 0$.

The conditions when a strongly risk-averse agent accepts the contract can be depicted by the shaded areas in Fig. 4.14, where $\eta = 2$. The two shaded areas with different grey scales represent conditions (4.24) and (4.26) under which the agent accepts the contract but responds differently. The lower bound function of the shaded areas (denoted by $w_0(p)$) represents the set of offers that give the agent zero expected utility rate. $w_0(p)$ is defined as follows:

$$w_0(p) = \begin{cases} p \text{ when } p \in (0, p_4] \\ 2\sqrt{(1+\eta)p\lambda} - \lambda \text{ when } p > p_4 \end{cases}$$

Since $\lim_{p \to p_4^-} w_0(p) = \lim_{p \to p_4^+} w_0(p) = p_4$, $w_0(p)$ is continuous everywhere over interval $p \in \mathbb{R}_+$. However since $\lim_{p \to p_4^-} dw_0(p)/dp = 1 \neq \sqrt{1+\eta}\left(\sqrt{1+\eta} + \sqrt{\eta}\right) = \lim_{p \to p_4^+} dw_0(p)/dp$, $w_0(p)$ is not differentiable at $p = p_4$.

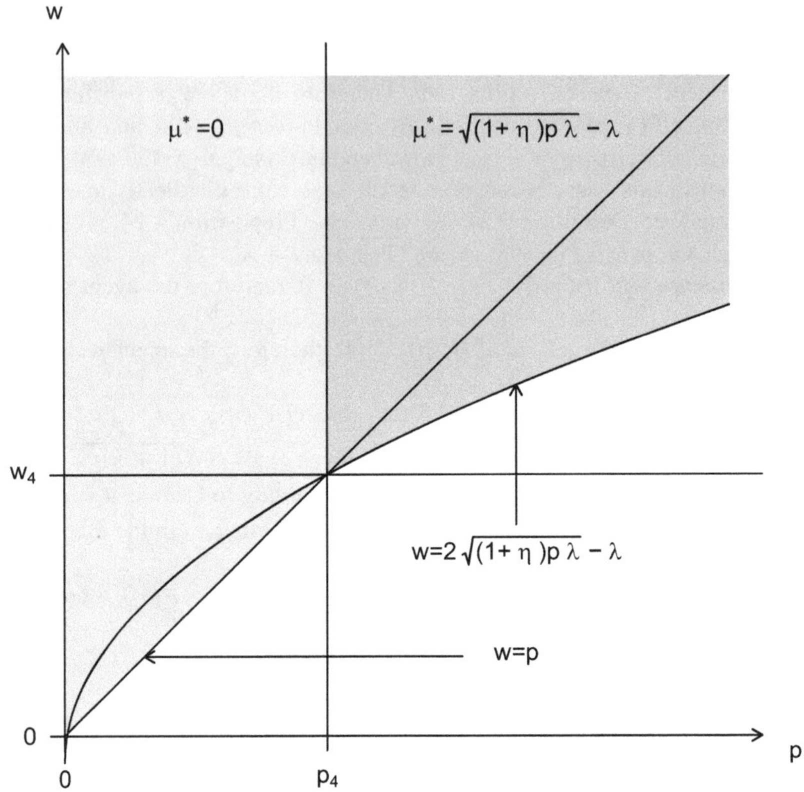

Fig. 4.14 Conditions when a strongly risk-averse agent accepts the contract with $\eta = 2$

4.2.1 *Sensitivity Analysis of a Strongly Risk-Averse Agent's Optimal Strategy*

Since the principal does not propose a contract that will be responded to with zero service capacity, therefore the only viable case is when the agent in response installs positive service capacity: $\mu^*(w,p) = \sqrt{(1+\eta)p\lambda} - \lambda$. The w is bounded below by $2\sqrt{(1+\eta)p\lambda} - \lambda = \eta pP(1) + pP(1) + \mu^*(w,p)$ (see (4.26)), with $\eta pP(1)$ representing the expected risk rate perceived by the agent and $pP(1)$ representing the expected penalty rate charged by the principal. It indicates that the agent has to be reimbursed for the expected risk rate, the expected penalty rate, and the cost of the optimal service capacity.

The optimal service capacity $\mu^*(w,p) = \sqrt{(1+\eta)p\lambda} - \lambda$ depends on the penalty rate p, the failure rate λ and the risk coefficient η. Its derivatives are:

$$\frac{\partial \mu^*}{\partial p} = \sqrt{\frac{(1+\eta)\lambda}{4p}} > 0,\ \frac{\partial \mu^*}{\partial \lambda} = \sqrt{\frac{(1+\eta)p}{4\lambda}} - 1 \text{ and } \frac{\partial \mu^*}{\partial \eta} = \sqrt{\frac{p\lambda}{4(1+\eta)}} > 0$$

The derivatives indicate that given λ and η the agent will increase the service capacity when the penalty rate increases. Note that $\sqrt{(1+\eta)p\lambda} - \lambda$, as a function of λ, increases when $(1+\eta)p/4 > \lambda$. From conditions (4.25) and (4.26) the agent installs service capacity $\sqrt{(1+\eta)p\lambda} - \lambda$ when $p \geq p_4$ and from Lemma 4.22 we have $p_4 > 4p_1$. Therefore we have $p > 4p_1 = 4\lambda/(1+\eta) \Rightarrow (1+\eta)p/4 > \lambda \Rightarrow \partial\mu^*/\partial\lambda > 0$. Thus, given p and η, the agent will increase the service capacity when the failure rate increases. Given the penalty rate and the failure rate, the agent will increase the service capacity as his risk-aversion increases.

The agent's optimal expected utility rate when installing capacity $\mu^*(w,p) = \sqrt{(1+\eta)p\lambda} - \lambda$ is $u_A^* \equiv u_A(\mu^*(w,p); w,p) = w - 2\sqrt{(1+\eta)p\lambda} + \lambda$, and it depends on w, p, η and λ. Note that $\partial u_A^*/\partial w = -1 < 0$, $\partial u_A^*/\partial p = -\sqrt{(1+\eta)\lambda/p} < 0$ and $\partial u_A^*/\partial \eta = -\sqrt{p\lambda/(1+\eta)} < 0$, indicating that the agent's optimal expected utility rate decreases with the compensation rate, the penalty rate and his risk intensity. Note that $\partial u_A^*/\partial\lambda = -\left(\sqrt{p} - \sqrt{p_1}\right)/\sqrt{p_1}$, and from Proposition 4.23 $p \geq p_4 > 4p_1 \Rightarrow \sqrt{p} - \sqrt{p_1} > 0$, therefore the agent's optimal expected utility rate also decreases with the failure rate.

Summary: Recall that a risk-neutral agent would accept a contract, install $\mu^*(w,p) = 0$ and receive $u(\mu^*(w,p); w,p) = w - p$ given the set of offers $\{(w,p) : p \in (0,\lambda], w \geq p\}$. Given the set of offers $\left\{(w,p) : p > \lambda, w \geq 2\sqrt{p\lambda} - \lambda\right\}$ he would accept the contract, install $\mu^*(w,p) = \sqrt{p\lambda} - \lambda$ and receive expected utility rate $u(\mu^*(w,p); w,p) = w - 2\sqrt{p\lambda} + \lambda$. By comparing the optimal solutions of a strongly risk-averse agent with that of a risk-neutral agent, two conclusions are drawn:

1. Given a λ, the principal must set a higher p in order to induce a strongly risk-averse agent to install a positive service capacity versus a risk-neutral agent ($p > \lambda$ for risk-neutral agent, $p > \left(1 + 2\eta + 2\sqrt{\eta(1+\eta)}\right)\lambda$ for strongly risk-averse agent).
2. With the same w and p, given that the agent accepts the contract and installs a positive service capacity, the expected utility rate of a strongly risk-averse agent decreases with respect to η since

$$u\left(\mu^*(w,p) = \sqrt{(1+\eta)p\lambda} - \lambda\right) = w - 2\sqrt{(1+\eta)p\lambda} + \lambda$$

$$\Rightarrow \frac{\partial u}{\partial \eta} = -\sqrt{\frac{p\lambda}{1+\eta}} < 0$$

Therefore a strongly risk-averse agent is always strictly worse off than a risk-neutral agent.

Compared to a weakly risk-averse agent, a strongly risk-averse agent has fewer options of positive optimal service capacities (he will never install $\mu^*(w,p) = \sqrt{(1-\eta)p\lambda} - \lambda$ when $\eta \in [4/5, 1)$) because the perceived risk rate is high enough such that the only reasonable choice is to invest more in service capacity to compensate for the risk.

4.2.2 Principal's Optimal Strategy

We now derive the principal's optimal strategy while anticipating the agent's optimal response $\mu^*(w,p)$. For that the principal solves the optimization problem:

$$\max_{w>0,p>0} \Pi_P(w,p;\mu^*(w,p)) = \max_{w>0,p>0} \left\{ \frac{r\mu^*(w,p)}{\lambda + \mu^*(w,p)} - w + \frac{p\lambda}{\lambda + \mu^*(w,p)} \right\} \tag{4.27}$$

and recovers the optimizing values: $(w^*, p^*) = \text{argmax}_{w>0,p>0} \Pi_P(w,p;\mu^*(w,p))$.

Before deriving the principal's optimal strategy, we reexamine the case when the principal offers $p = p_4$ and $w \geq w_4$, under which the agent is indifferent regarding two different service capacities which however effect the principal differently. Since any selected solution $((w,p),\mu)$ has to be an admissible solution (see Definition 2.3) we test the solutions' membership in Proposition 4.24.

Proposition 4.24. *Suppose a strongly risk-averse agent. Assume that the principal's possible offers are constrained to set* $\{(w,p) : p = p_4, w \geq w_4\}$.

(a) *If* $r \in (0, p_4]$, *then the principal does not propose a contract.*
(b) *If* $r > p_4$, *the agent installs* $\sqrt{(1+\eta)p_4\lambda} - \lambda$ *if offered a contract.*

Proof. For $w \geq w_4$ we have $\partial\Pi_P(w,p_4;\mu)/\partial\mu = (r - p_4)\lambda/(\lambda+\mu)^2$. Define $\mu_L \equiv 0$ and $\mu_H \equiv \sqrt{(1+\eta)p_4\lambda} - \lambda$ and note that $\mu_H > \mu_L$. If $r \in (0, p_4)$, then $\partial\Pi_P/\partial\mu < 0$, therefore $((w,p_4),\mu_L) \succeq ((w,p_4),\mu_H)$ and the agent would install μ_L if offered a contract since $((w,p_4),\mu_H)$ is not an admissible solution. However in such case the principal's expected profit rate is $\Pi_P(w,p_4;\mu_L) = -w + p_4 \leq 0$, therefore the principal would not propose a contract. If $r = p_4$, then $\partial\Pi_P/\partial\mu = 0$, therefore the agent installs either μ_L or μ_H if offered a contract. However in such case the principal's expected profit rate is $\Pi_P(w,p_4;\mu_L) = \Pi_P(w,p_4;\mu_H) = -w + p_4 \leq 0$, therefore the principal would not propose a contract. If $r > p_4$, then $\partial\Pi_P/\partial\mu > 0$ and $((w,p_4),\mu_H) \succeq ((w,p_4),\mu_L)$. If the principal offers a contract (where the conditions will be discussed in detail in Theorem 4.27 that follows), then by Definition 2.3 only μ_H leads to admissible solutions. □

Notation:

$$r_4 \equiv \left(1 + 2\eta + 2\sqrt{\eta(1+\eta)}\right)(1+2\eta)\lambda = (1+2\eta)p_4 \tag{4.28}$$

r_4 is a function of λ and η however we suppress the parameters (λ, η).

Next we state several technical lemmas (see proofs in the Appendix).

Lemma 4.25. *Consider* $\max_{x \geq \sqrt{p_4}} f(x)$ *where* $f(x) = r + \lambda - \sqrt{p_1}\,((1+2\eta)x + r/x)$ *and denote* $x^* = \operatorname{argmax}_{x \geq \sqrt{p_4}} f(x)$. *The solutions to this optimization problem are*

(a) $x^* = \sqrt{p_4}$ *if* $r \in (0, r_4]$.
(b) $x^* = \sqrt{r/(1+2\eta)}$ *if* $r > r_4$.

Lemma 4.26. *Let* $\eta > 0$ *and* $\lambda > 0$, *then* $(1+2\eta)p_4 > \left(1 + 3\eta + 2\sqrt{\eta(1+2\eta)}\right)\lambda/(1+\eta)$.

The principal's optimal strategy is derived in Theorem 4.27. Recall that Proposition 4.23 describes the agent's optimal response to each pair of compensation rate and penalty rate $(w, p) \in \mathbb{R}_+^2$. Since the principal will not propose a contract that is going to be rejected by a strongly risk-averse (SRA) agent, therefore Theorem 4.27 only considers pairs $(w, p) \in \mathbb{R}_+^2$ such that the agent receives a non-negative expected utility rate. Define

$$\begin{aligned}
\mathfrak{D}_{(4.24)} &\equiv \{(w,p) \text{ that satisfies (4.24) when } \eta \geq 4/5\} \\
\mathfrak{D}_{(4.25)} &\equiv \{(w,p) \text{ that satisfies (4.25) when } \eta \geq 4/5\} \\
\mathfrak{D}_{(4.26)} &\equiv \{(w,p) \text{ that satisfies (4.26) when } \eta \geq 4/5\} \\
\mathfrak{D}_{\text{SRA}} &\equiv \mathfrak{D}_{(4.24)} \cup \mathfrak{D}_{(4.25)} \cup \mathfrak{D}_{(4.26)}
\end{aligned} \tag{4.29}$$

Theorem 4.27. *Given a strongly risk-averse agent and* $(w, p) \in \mathfrak{D}_{\text{SRA}}$.

(a) *If* $r \in (0, p_4]$, *then the principal does not propose a contract.*
(b) *If* $r \in (p_4, r_4]$, *then the principal's offer and the capacity installed by the agent are*

$$(w^*, p^*) = (w_4, p_4) \text{ and } \mu^*(w^*, p^*) = \sqrt{(1+\eta)p_4\lambda} - \lambda \tag{4.30}$$

and the principal's expected profit rate is

$$\Pi_P(w^*, p^*; \mu^*(w^*, p^*)) = r + \lambda - \sqrt{p_1}\left((1+2\eta)\sqrt{p_4} + \frac{r}{\sqrt{p_4}}\right) \tag{4.31}$$

(c) *If $r > r_4$, then the principal's offer and the capacity installed by the agent are*

$$(w^*, p^*) = \left(2\sqrt{\frac{(1+\eta)r\lambda}{1+2\eta}} - \lambda, \frac{r}{1+2\eta}\right) \text{ and } \mu^*(w^*, p^*) = \sqrt{\frac{(1+\eta)r\lambda}{1+2\eta}} - \lambda \tag{4.32}$$

and the principal's expected profit rate is

$$\Pi_P(w^*, p^*; \mu^*(w^*, p^*)) = r - 2\sqrt{\frac{(1+2\eta)r\lambda}{1+\eta}} + \lambda \tag{4.33}$$

Proof. The structure of the proof is depicted in Fig. 4.15.

Case $(w, p) \in \mathfrak{D}_{(4.24)} \cup \mathfrak{D}_{(4.25)}$: According to Proposition 4.23 part (a), if $(w, p) \in \mathfrak{D}_{(4.24)}$, then in case the principal makes an offer, the agent accepts the contract but does not install any service capacity. Since $\partial\Pi_P/\partial w = -1 < 0$, therefore $w^* = p$ and from Eq. (3.3) $\Pi_P(w^*, p; \mu^*(w^*, p)) = -w^* + p = -p + p = 0$. According to Propositions 4.23 part (b) and 4.24, if $(w, p) \in \mathfrak{D}_{(4.25)}$ (which implies $p = p_4$), then the principal does not propose a contract if $r \in (0, p_4]$, or installs $\sqrt{(1+\eta)p_4\lambda} - \lambda$ in case the principal makes an offer when $r > p_4$. Since $\partial\Pi_P/\partial w = -1 < 0$, therefore $w^* = w_4$. From Proposition 4.23 part (b), if the principal offers a contract with $(w, p) = (w_4, p_4)$, then the agent installs either $\mu^*(w_4, p_4) = 0$ or $\mu^*(w_4, p_4) = \sqrt{(1+\eta)p_4\lambda} - \lambda$. Denote the principal's expected profit rate when $(w, p) = (w_4, p_4)$ and $\mu^*(w, p) = 0$ by $\Pi_P^L(p_4)$, and denote the principal's expected profit rate when $(w, p) = (w_4, p_4)$ and $\mu^*(w, p) = \sqrt{(1+\eta)p_4\lambda} - \lambda$ by $\Pi_P^H(p_4)$. By plugging the value of w, p and μ into Eq. (3.3):

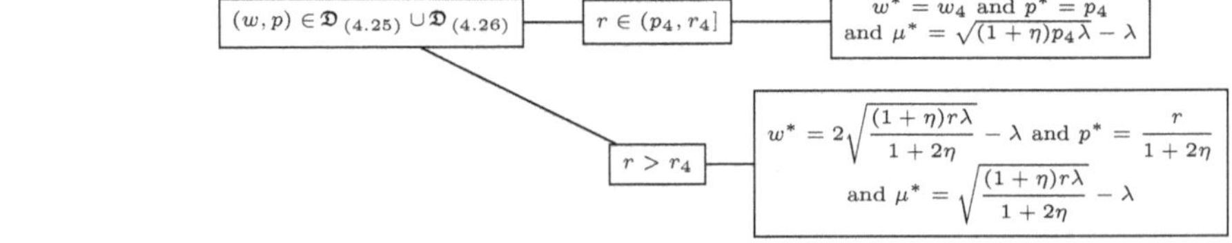

Fig. 4.15 Structure of the proof for Theorem 4.27

$$\Pi_P^L(p_4) = -w_4 + p_4 = 0 \tag{4.34}$$

$$\Pi_P^H(p_4) = r + \lambda - \sqrt{p_1}\left((1+2\eta)\sqrt{p_4} + \frac{r}{\sqrt{p_4}}\right) = \left(\frac{\sqrt{p_4} - \sqrt{p_1}}{\sqrt{p_4}}\right)(r - p_4) \tag{4.35}$$

In such case the principal's optimization problem is $\max_{p \in (0,p_4]} \Pi_P(w^*, p; \mu^*(w^*, p))$ where:

$$\Pi_P(w^*, p; \mu^*(w^*, p)) = \begin{cases} 0 \text{ for } p \in (0, p_4) \\ \max\left\{\Pi_P^L(p_4), \Pi_P^H(p_4)\right\} \text{ for } p = p_4 \end{cases}$$

Subcase $r \in (0, p_4]$: By Proposition 4.24 part (a), the principal does not offer a contract.

Subcase $r > p_4$: According to Lemma 4.25 part (b), $p^* = p_4$ and according to Proposition 4.24 part (b) the principal's expected profit rate $\Pi_P(w^*, p^*; \mu^*(w^*, p^*)) = \Pi_P^H(p_4) > \Pi_P^L(p_4) = 0$. Thus the principal proposes a contract with $w^* = w_4$ and $p^* = p_4$ that induces the agent to install $\mu^*(w^*, p^*) = \sqrt{(1+\eta)p_4\lambda} - \lambda$.

<u>**Case $(w, p) \in \mathfrak{D}_{(4.25)} \cup \mathfrak{D}_{(4.26)}$:**</u> According to Proposition 4.23 part (c), if $(w, p) \in \mathfrak{D}_{(4.26)}$, then in case the principal makes an offer, the agent accepts the contract and installs $\sqrt{(1+\eta)p\lambda} - \lambda$. Since $\partial\Pi_P/\partial w = -1 < 0$, therefore $w^* = 2\sqrt{(1+\eta)p\lambda} - \lambda$. According to Propositions 4.23 part (b) and 4.24, if $(w, p) \in \mathfrak{D}_{(4.25)}$ (which implies $p = p_4$), then the principal does not propose a contract if $r \in (0, p_4]$, or installs $\sqrt{(1+\eta)p_4\lambda} - \lambda$ in case the principal makes an offer when $r > p_4$. Since $\partial\Pi_P/\partial w = -1 < 0$, therefore $w^* = w_4$. From Proposition 4.23 part (b), if the principal offers a contract with $(w, p) = (w_4, p_4)$, then the agent installs either $\mu^*(w_4, p_4) = 0$ or $\mu^*(w_4, p_4) = \sqrt{(1+\eta)p_4\lambda} - \lambda$. The principal's optimization problem is $\max_{p \geq p_4} \Pi_P(w^*, p; \mu^*(w^*, p))$ where:

$$\Pi_P(w^*, p; \mu^*(w^*, p)) = \begin{cases} \max\left\{\Pi_P^L(p_4), \Pi_P^H(p_4)\right\}, \text{ for } p = p_4 \\ r + \lambda - \sqrt{p_1}\left((1+2\eta)\sqrt{p} + \dfrac{r}{\sqrt{p}}\right), \text{ for } p > p_4 \end{cases}$$

Define $x \equiv \sqrt{p}$, the expression $r + \lambda - \sqrt{p_1}\left((1+2\eta)\sqrt{p} + r/\sqrt{p}\right)$ can be restated as $f(x) = r + \lambda - \sqrt{p_1}\left((1+2\eta)x + r/x\right)$. Maximizing $f(x)$ with respect for $x \geq \sqrt{p_4}$ is equivalent to maximizing $r + \lambda - \sqrt{p_1}\left((1+2\eta)\sqrt{p} + r/\sqrt{p}\right)$ for $p \geq p_4$ in the sense that

$$\underset{p \geq p_4}{\operatorname{argmax}}\left\{r + \lambda - \sqrt{p_1}\left((1+2\eta)\sqrt{p} + \frac{r}{\sqrt{p}}\right)\right\} = \left(\underset{x \geq \sqrt{p_4}}{\operatorname{argmax}} f(x)\right)^2$$

Since $r_4 = (1+2\eta)p_4 > p_4$, we examine the following subcases.

Subcase $r \in (0, p_4]$: By Proposition 4.24 part (a), the principal does not propose a contract.

Subcase $r \in (p_4, r_4]$: According to Lemma 4.25 part (a), $p^* = p_4$. According to Proposition 4.24 part (b), $\Pi_P(w^*, p^*; \mu^*(w^*, p^*)) = \Pi_P^H(p_4) > \Pi_P^L(p_4) = 0$. Therefore the principal proposes a contract with $w^* = w_4$ and $p^* = p_4$ that induces the agent to install $\mu^*(w^*, p^*) = \sqrt{(1+\eta)p_4\lambda} - \lambda$.

Subcase $r > r_4$: According to Lemma 4.25 part (b), $p^* = r/(1+2\eta)$ and the principal's expected profit rate is $\Pi_P(w^*, p^*; \mu^*(w^*, p^*)) = r - 2\sqrt{(1+2\eta)r\lambda/(1+\eta)} + \lambda$. According to Lemmas 4.14 and 4.26 $\Pi_P(w^*, p^*; \mu^*(w^*, p^*)) > 0$, therefore the principal proposes a contract with $w^* = 2\sqrt{(1+\eta)r\lambda/(1+2\eta)} - \lambda$ and $p^* = r/(1+2\eta)$ that induces the agent to install service capacity $\mu^*(w^*, p^*) = \sqrt{(1+\eta)r\lambda/(1+2\eta)} - \lambda$.

To summarize, if $r \in (0, p_4]$, then the principal does not propose a contract. This case corresponds to Theorem 4.27 (a). If $r \in (p_4, r_4]$, then the principal offers $(w^*, p^*) = (w_4, p_4)$ and the agent installs capacity $\mu^*(w^*, p^*) = \sqrt{(1+\eta)p_4\lambda} - \lambda$. This case corresponds to Theorem 4.27 (b). Finally if $r > r_4$, then according to Lemma 4.25 part (b), the principal offers $(w^*, p^*) = \left(2\sqrt{(1+\eta)r\lambda/(1+2\eta)} - \lambda, r/(1+2\eta)\right)$ and the agent installs capacity $\mu^*(w^*, p^*) = \sqrt{(1+\eta)r\lambda/(1+2\eta)} - \lambda$. This case corresponds to Theorem 4.27 (c). □

Theorem 4.27 indicates that the existence of a beneficial contract for strongly risk-averse agent is determined exogenously by the market (the revenue rate r), the nature of the equipment (the failure rate λ) and the nature of the agent (the risk coefficient η).

4.3 Risk-Averse Agent: A Summary

Recall the definition of p_2, p_3, p_4, r_2, r_3 and r_4 from (4.5), (4.12), (4.23) and (4.28). The conditions that a principal makes offers to a risk-averse agent is depicted by the shaded areas in Fig. 4.16. The horizontal axis represents the agent's risk coefficient, and the vertical axis represents the revenue rate generated by the principal's unit, which is exogenously determined by the market. The principal makes different offers to the agent when (r, η) is in the five shaded areas with different gray scales. We define

$$p_{34} \equiv \begin{cases} p_3 \text{ for } \eta \in (0, 4/5) \\ p_4 \text{ for } \eta \geq 4/5 \end{cases} \quad \text{and } r_{34} \equiv \begin{cases} r_3 \text{ for } \eta \in (0, 4/5) \\ r_4 \text{ for } \eta \geq 4/5 \end{cases}$$

Note that $\lim_{\eta \to (4/5)^-} r_{34} = 13\lambda = \lim_{\eta \to (4/5)^+} r_{34}$, and note that $\lim_{\eta \to (4/5)^-} \partial r_{34}/\partial\eta = 125\lambda/6 = \lim_{\eta \to (4/5)^+} \partial r_{34}/\partial\eta$, therefore r_{34} is continuous and differentiable everywhere over $\mathbb{R}_+$. Since $\lim_{\eta \to (4/5)^-} p_{34} = 5\lambda = \lim_{\eta \to (4/5)^+} p_{34}$ and $\lim_{\eta \to (4/5)^-} \partial p_{34}/\partial\eta = 25\lambda/6 = \lim_{\eta \to (4/5)^+} \partial p_{34}/\partial\eta$, therefore p_{34} is

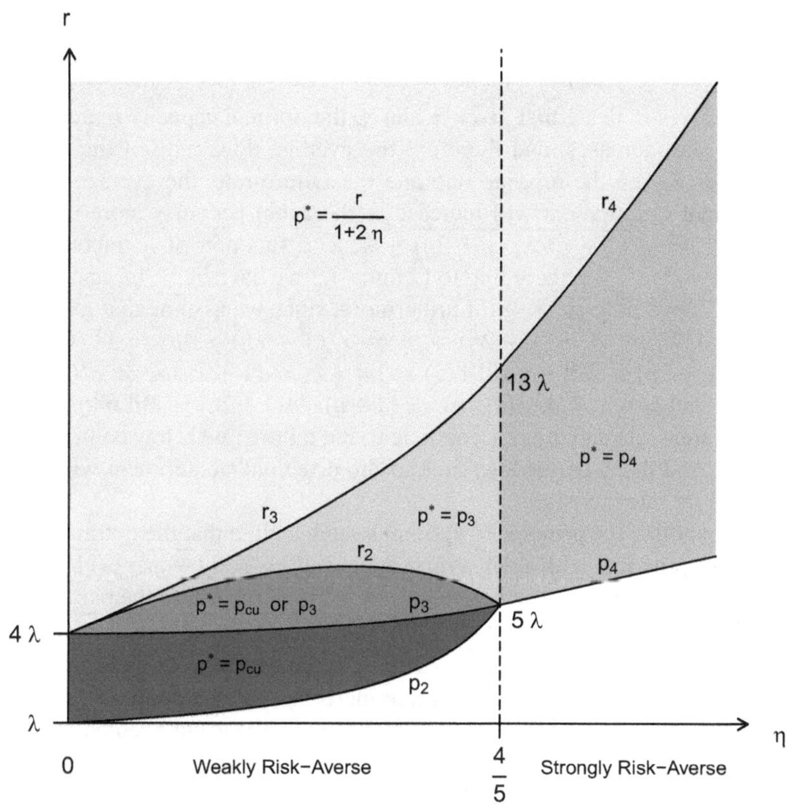

Fig. 4.16 Conditions when a risk-neutral principal makes offers to a risk-averse agent

continuous and differentiable everywhere over $\mathbb{R}_+$ as well. Furthermore, note that $\lim_{\eta\to 0^+} p_3 = \lim_{\eta\to 0^+} r_2 = \lim_{\eta\to 0^+} r_3 = 4\lambda$ and $\lim_{\eta\to(4/5)^-} r_2 = \lim_{\eta\to(4/5)^-} p_2 = 5\lambda$.

4.3.1 *Sensitivity Analysis of Optimal Strategies in High Revenue Industry*

The revenue rate r is determined exogenously by the market, and consider $r > r_{34}$ (high revenue rate). Equations (4.21) and (4.32) are the second-best solutions when the agent is weakly and strongly risk-averse respectively, and they have the same functional form.

The risk-averse agent's optimal strategy is examined first. Note that the optimal service capacity of a risk-averse agent ($\mu^* = \sqrt{(1+\eta)r\lambda/(1+2\eta)} - \lambda$) is a function of r, λ, and η. The derivatives of μ^* with respect to the parameters are $\partial\mu^*/\partial r = \sqrt{(1+\eta)\lambda}/\left(2\sqrt{(1+2\eta)r}\right) > 0$, $\partial\mu^*/\partial\lambda = \sqrt{(1+\eta)r}/\left(2\sqrt{(1+2\eta)\lambda}\right)-1$ and $\partial\mu^*/\partial\eta = -\sqrt{r\lambda}/\left(2\sqrt{(1+\eta)(1+2\eta)^3}\right) < 0$. The derivatives indicate that given λ and η, the optimal capacity increases when the revenue rate increases, and therefore the average downtime of the principal's unit decreases. Given the revenue rate and the failure rate, the average downtime of the principal's equipment will increase as the agent becomes more risk-averse. Note that $\mu^* = \sqrt{(1+\eta)r\lambda/(1+2\eta)} - \lambda$, as a function of λ, increases when $(1+\eta)r/4(1+2\eta) > \lambda$. According to Lemma 4.3 we have $p_3 > 4p_1$ and according to Lemma 4.22 we have $p_4 > 4p_1$. Furthermore, since we assume that $r > r_{34}$, then $r > r_3 = (1+2\eta)p_3 \Rightarrow r/(1+2\eta) > p_3 \Rightarrow r/(1+2\eta) > 4p_1 = 4\lambda/(1+\eta) \Rightarrow (1+\eta)r/4(1+2\eta) > \lambda$ if $\eta \in (0, 4/5)$ and $r > r_4 = (1+2\eta)p_4 \Rightarrow r/(1+2\eta) > p_4 \Rightarrow r/(1+2\eta) > 4p_1 = 4\lambda/(1+\eta) \Rightarrow (1+\eta)r/4(1+2\eta) > \lambda$ if $\eta \geq 4/5$. Thus, given the revenue rate and the risk coefficient, the failure rate is low compared to the revenue rate, and the average downtime of the principal's equipment will decrease when the failure rate increases.

Next we examine the principal's optimal strategy. Note that the optimal compensation rate of a principal with a risk-averse agent ($w^* = 2\sqrt{(1+\eta)r\lambda/(1+2\eta)}-\lambda$) is a function of r, λ, and η. The derivatives of w^* with respect to the parameters are $\partial w^*/\partial r = \sqrt{(1+\eta)\lambda/(1+2\eta)r} > 0$, $\partial w^*/\partial\lambda = \sqrt{(1+\eta)r/(1+2\eta)\lambda} - 1$ and finally $\partial w^*/\partial\eta = -\sqrt{r\lambda/(1+\eta)(1+2\eta)^3} < 0$. The derivatives indicate that given the λ and η, the optimal compensation rate increases with respect to r. Given the r and the λ, the optimal compensation rate decreases as the agent becomes more risk-averse. Note that $w^* = 2\sqrt{(1+\eta)r\lambda/(1+2\eta)} - \lambda$, as a function of λ, increases when $(1+\eta)r/(1+2\eta) > \lambda$. According to Lemma 4.3 we have $p_3 > 4p_1 > p_1$ and according to Lemma 4.22 we have $p_4 > 4p_1 > p_1$. Furthermore, since we assume that $r > r_{34}$, then $r > r_3 = (1+2\eta)p_3 \Rightarrow r/(1+2\eta) > p_3 \Rightarrow r/(1+2\eta) > p_1 = \lambda/(1+\eta) \Rightarrow (1+\eta)r/(1+2\eta) > \lambda$ if $\eta \in (0, 4/5)$ and $r > r_4 = (1+2\eta)p_4 \Rightarrow r/(1+2\eta) > p_4 \Rightarrow r/(1+2\eta) > p_1 = \lambda/(1+\eta) \Rightarrow (1+\eta)r/(1+2\eta) > \lambda$ if $\eta \geq 4/5$. Therefore the failure rate is low compared to the revenue rate ($(1+\eta)r/(1+2\eta) > \lambda \Rightarrow \partial w^*/\partial\lambda > 0$), indicating that the w^* increases with respect to the failure rate.

The principal's optimal p^* given a risk-averse agent ($p^* = r/(1+2\eta)$) is a function of r and η. Note that p^* is independent of the failure rate λ under the assumption that the revenue rate is sufficiently high compared to the failure rate. The derivatives of p^* with respect to the parameters are $\partial p^*/\partial r = 1/(1+2\eta) > 0$ and $\partial p^*/\partial\eta = -2r/(1+2\eta)^2 < 0$. The derivatives indicate that given the risk η, the optimal penalty p^* increases with respect to r, and given r, the p^* decreases with respect to η.

The principal's optimal expected profit rate given a risk-averse agent

$$\Pi_P^* \equiv \Pi_P(w^*, p^*; \mu^*(w^*, p^*)) = r - 2\sqrt{(1+2\eta)r\lambda/(1+\eta)} + \lambda$$

is a function of r, λ, and η. The derivatives of Π_P^* with respect to these parameters are $\partial\Pi_P^*/\partial r = 1 - \sqrt{(1+2\eta)\lambda/(1+\eta)r}$, $\partial\Pi_P^*/\partial\lambda = 1 - \sqrt{(1+2\eta)r/(1+\eta)\lambda}$, $\partial\Pi_P^*/\partial\eta = -\sqrt{r\lambda/(1+2\eta)(1+\eta)^3} < 0$. The derivatives indicate that given r and λ, the principal's optimal expected profit rate decreases as the agent becomes more risk-averse. Note that $\Pi_P^* = r - 2\sqrt{(1+2\eta)r\lambda/(1+\eta)} + \lambda$, as a function of λ, decreases when $(1+2\eta)r/(1+\eta) > \lambda$, and as a function of r, increases when $r > (1+2\eta)\lambda/(1+\eta)$. According to Lemma 4.3 we have $p_3 > 4p_1 > p_1$ and according to Lemma 4.22 we have $p_4 > 4p_1 > p_1$. Furthermore, since we assume that $r > r_{34}$, then $r > r_3 = (1+2\eta)p_3 \Rightarrow r/(1+2\eta) > p_3 \Rightarrow r/(1+2\eta) > p_1 = \lambda/(1+\eta)$ if $\eta \in (0, 4/5)$ and $r > r_4 = (1+2\eta)p_4 \Rightarrow r/(1+2\eta) > p_4 \Rightarrow r/(1+2\eta) > p_1 = \lambda/(1+\eta)$ if $\eta \geq 4/5$. Therefore given a λ and an η, the revenue rate is high compared to the failure rate ($r > (1+2\eta)\lambda/(1+\eta) \Rightarrow \partial\Pi_P^*/\partial r > 0$), thus the principal's optimal expected profit rate increases with respect to the revenue rate. Note that since $\eta > 0$, therefore $r > (1+2\eta)\lambda/(1+\eta) > (1+\eta)\lambda/(1+2\eta) \Rightarrow (1+2\eta)r/(1+\eta) > \lambda$, which implies that given an r and η, the failure rate is low compared to the revenue rate ($(1+2\eta)r/(1+\eta) > \lambda \Rightarrow \partial\Pi_P^*/\partial\lambda < 0$), therefore the principal's optimal expected profit rate decreases with respect to λ.

4.3.2 The Second-Best Solution in High Revenue Industry

By comparing the second-best solution given a risk-averse agent ((4.21) and (4.32)) with the second-best given a risk-neutral agent when $r > r_{34}$, four conclusions are drawn.

1. The optimal w^* and the optimal p^* decrease when the agent is risk-averse versus risk-neutral agent ($w^* : 2\sqrt{r\lambda} - \lambda > 2\sqrt{(1+\eta)r\lambda/(1+2\eta)} - \lambda$ and $p^* : r > r/(1+2\eta)$). It indicates that the risk adds an incentive for the agent to install a higher service capacity by coupling it to the penalty charge collected by the principal.
2. The principal is worse off with a risk-averse agent than a risk-neutral agent ($r - 2\sqrt{r\lambda} + \lambda > r - 2\sqrt{(1+2\eta)r\lambda/(1+\eta)} + \lambda$), as well as with an agent whose action is contractible (recall that the principal receives the same expected profit rate with a risk-neutral agent in first-best and second-best setting). This conclusion is consistent with Proposition 3 part (ii) in Harris and Raviv (1978). The principal's loss can be explained as follows: On one hand, the decrease in the agent's optimal capacity when risk-averse reduces the revenue performance of the principal's unit. At the same time, the monetary equivalency of the risk perceived by the agent is not channeled to the principal, although from the agent's perspective it serves as part of the penalty charge.
3. The μ^* of a risk-averse agent is strictly less than that of a risk-neutral agent ($\sqrt{r\lambda} - \lambda > \sqrt{(1+\eta)r\lambda/(1+2\eta)} - \lambda$). Recall that when the agent is risk-neutral, the μ^* in the second-best solution is the same as that in the first-best solution, indicating that the unobservability of the agent's service capacity does

not contribute to the decrease of the optimal service capacity. When the agent is risk-averse, he compensates for the risk he bears by reducing μ.

4. Given the compensation rate and penalty rate, both weakly and strongly risk-averse agents are worse off compared to a risk-neutral agent.

To summarize, for a principal with high revenue generating unit, agent's risk-aversion reduces the efficiency of the contract (compared to the first-best contract), and therefore it reduces the social welfare.

Chapter 5
Risk-Seeking Agent

In previous section we represented agent's perceived risk by a measure that reflects the dispersion of his revenue stream. Although the dispersion of possible outcomes has been widely used as the measure of risk (Pratt 1964; Rothschild and Stiglitz 1970; Stiglitz 1974; Levy 1992; Fukunaga and Huffman 2009; Lewis and Bajari 2014) it fails to capture observable behavior in risky settings. In this section we extend our principal-agent analysis to risk-seeking agent. We note that there is an ongoing evaluation of risk attitudes in an attempt to explain peoples' behavior when faced with risky choices. For instance Prospect Theory claims to offer a better model that covers discrepancies observed elsewhere (Kahneman and Tversky 1979; Tversky and Kahneman 1992). Prospect Theory claims that people are less sensitive to the variation of the probability of outcomes compared to the expectation, and losses loom larger than gains. Furthermore, empirical evidences indicates that decision makers prefer expressions of risk in terms of the expected value at stake, and they appear to be risk-averse when dealing with a risky alternative whose possible outcomes are generally good and tend to be risk-seeking when dealing with a risky alternative whose possible outcomes are generally poor (March and Shapira 1987; Filiz-Ozbay et al. 2013).

In our principal-agent setting with a risk-seeking agent we propose that an agent perceives a greater loss when he is charged a larger penalty rate for each unit of downtime and also when the probability of being in the failed state goes up. The agent's penalty rate at any point of time can be modeled as pB where B is a Bernoulli random variable that takes value 0 with probability $P(0) = \mu/(\lambda + \mu)$ and value 1 with probability $P(1) = \lambda/(\lambda + \mu)$. For simplicity denote momentarily $a \equiv P(1)$. In this section we adopt the following risk measure:

$$r(a) \equiv p\left(a - \frac{1}{2}\right)_+ \quad \text{for } a \in [0, 1]$$

S. Zeng, M. Dror, *Formulating Principal-Agent Service Contracts for a Revenue Generating Unit*, SpringerBriefs in Operations Management,
DOI 10.1007/978-3-319-18672-6_5

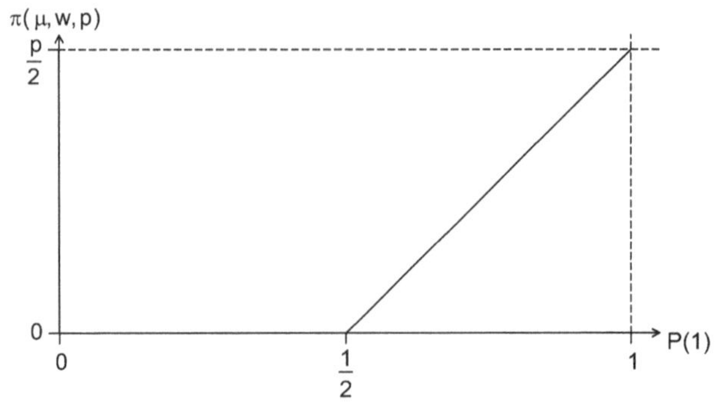

Fig. 5.1 $\pi(\mu, w, p)$ as a function of $P(1)$ when $\eta = -1$

We note that $R(pB) \equiv r(a)$ satisfies the properties of *monotonicity*, *sub-additivity* and *positive homogeneity* of a coherent risk measure but fails to satisfy the property of *translation invariance*, since $R(pB)$ is independent of the expectation of pB (Artzner et al. 1999).

Risk premium of a risk-seeking agent is the \$ value considered by the agent as extra gains to his revenue stream. As a consequence, just for the risk-seeking agent we modify the risk premium defined earlier in (4.1), in a manner that reflects the expected amount at stake instead of the dispersion of the revenue stream:

$$\pi(\mu, w, p) \equiv -\eta p\left(P(1) - \frac{1}{2}\right)_+ = -\eta p\left(\frac{\lambda}{\lambda + \mu} - \frac{1}{2}\right)_+ \tag{5.1}$$

Note that for risk-seeking agent $\eta < 0 \Rightarrow \pi(\mu, w, p) \geq 0$, and adding such a risk premium to a risk-neutral agent's expected utility rate (as in (5.2)) implies risk-seeking. Figure 5.1 depicts $\pi(\mu, w, p)$ as a function of $P(1)$ for $\eta = -1$.

The representation of the risk premium in (5.1) is consistent with the properties of risk in the Prospect Theory (Kahneman and Tversky 1979; Tversky and Kahneman 1992) and the empirical findings in (March and Shapira 1987): The risk premium is zero when $P(1)$ is lower than $1/2$. The risk premium increases with $P(1)$ linearly when $P(1)$ exceeds one half, and reaches its peak when $P(1) = 1$.

Denote $\overline{\eta} \equiv -\eta > 0$. Modifying (3.2), the risk-seeking agent's expected utility rate is:

$$u_A(\mu; w, p) = \left(w - \frac{p\lambda}{\lambda + \mu} - \mu + \overline{\eta} p\left(\frac{\lambda}{\lambda + \mu} - \frac{1}{2}\right)_+\right)_+ \quad \text{for } w > 0, p > 0, \mu \geq 0 \tag{5.2}$$

Since the analysis is different for $\overline{\eta} \in (0, 8/9)$, $\overline{\eta} \in [8/9, 2)$ and $\overline{\eta} \geq 2$, therefore, when $\overline{\eta} \in (0, 8/9)$ we consider the agent as *weakly risk-seeking*, when $\overline{\eta} \in [8/9, 2)$ we consider the agent as *moderately risk-seeking*, and when $\overline{\eta} \geq 2$ we consider the agent as *strongly risk-seeking*. We assume, say for historic reasons, that both the agent and the principal know not only the agent's type as risk-seeking but also the value of $\overline{\eta}$.

The expression for the principal's expected revenue rate $\Pi_P(w, p; \mu)$ remains the same as (3.3).

Before examining the details of the optimal contracts we discuss a potential case of the agent compensating the principal at times during the contract. Such occurrence of utility transfer from an risk-seeking agent to the principal can have one of two forms: either the compensation rate is non-positive ($w \leq 0$), or the principal is guaranteed a positive expected revenue rate even with her unit in the failed state forever ($-w + p > 0$ if $\mu = 0$). Under our setting of undetermined contract horizon it is unrealistic to accept that the agent might compensate the principal when the unit is forever in the failed state. Therefore the occurrence of a non-positive compensation rate ($w \leq 0$) has be ruled out in the definition of the Strategy Set (Definition 2.1). Nevertheless, the possibility of the principal receiving a positive expected revenue rate with a failed unit has to be considered. Therefore we extend the definition of the *Set of Admissible Solutions* (Definition 2.3) as follows.

Definition 5.1 (*Set of Admissible Solutions*). *The* ***set of admissible solutions*** *for the principal-agent problem* $\mathfrak{P}$ *is the set* $\mathfrak{s}(\mathfrak{P})$ *of all strategies* $((w,p),\mu) \in \mathfrak{S}(\mathfrak{P})$ *for which:*

(a) $\nexists\ ((w',p'),\mu') \in \mathfrak{S}(\mathfrak{P})$ *such that* $((w',p'),\mu') \succeq ((w,p),\mu)$ *– there is no other strategy that weakly dominates* $((w,p),\mu)$.
(b) $\Pi_P(w,p;\mu) > \underline{\Pi_P}$ *and* $u_A(\mu; w,p) \geq \underline{u_A}$.
(c) *If* $\mu = 0$, *then* $w \geq p$.

We denote the part inside the brackets in Eq. (5.2) as

$$u(\mu) \equiv \begin{cases} w - \dfrac{\overline{\eta}p}{2} - \dfrac{(1-\overline{\eta})p\lambda}{\lambda+\mu} - \mu, \ \mu \in [0,\lambda] \\ w - \dfrac{p\lambda}{\lambda+\mu} - \mu, \ \mu > \lambda \end{cases} \tag{5.3}$$

Note that $u(\mu)$ is differentiable everywhere for $\mu \geq 0$ except at $\mu = \lambda$. When $\mu \in [0,\lambda)$:

$$\frac{du(\mu)}{d\mu} = \frac{(1-\overline{\eta})p\lambda}{(\lambda+\mu)^2} - 1, \quad \lim_{\mu\to 0+} \frac{du(\mu)}{d\mu} = \frac{1-\overline{\eta}}{\lambda}\left(p - \frac{\lambda}{1-\overline{\eta}}\right)$$

$$\lim_{\mu\to\lambda-} \frac{du(\mu)}{d\mu} = \frac{1-\overline{\eta}}{4\lambda}\left(p - \frac{4\lambda}{1-\overline{\eta}}\right), \quad \frac{d^2u(\mu)}{d\mu^2} = -\frac{2(1-\overline{\eta})p\lambda}{(\lambda+\mu)^3}$$

and when $\mu > \lambda$:

$$\frac{du(\mu)}{d\mu} = \frac{p\lambda}{(\lambda+\mu)^2} - 1, \quad \lim_{\mu\to\lambda+} \frac{du(\mu)}{d\mu} = \frac{p-4\lambda}{4\lambda}$$

$$\lim_{\mu\to+\infty} \frac{du(\mu)}{d\mu} = -1 < 0 \text{ and } \frac{d^2u(\mu)}{d\mu^2} = -\frac{2p\lambda}{(\lambda+\mu)^3} < 0$$

The positivity or negativity of the above derivatives indicate the direction of monotonicity and the concavity/convexity of the function $u(\mu)$ over $[0,\lambda)$ and $(\lambda,+\infty)$. Table 5.1 summarizes these indicators for various regions of the space $\mathbb{R}^2_+$ for pairs of $(\overline{\eta}, p)$. In the table $u_\mu(\cdot) = \lim_{\mu\to\cdot} du/d\mu$, and $u_\mu(\cdot^+)$ represents the limit of $u_\mu(\mu)$ as μ approaches $(\cdot)$ from above, and similarly $u_\mu(\cdot^-)$ represents the limit of $u_\mu(\mu)$ as μ approaches $(\cdot)$ from below.

5.1 Optimal Strategies for the Weakly Risk-Seeking Agent

Note that agent's expected utility rate (see (5.2)) increases and principal's expected profit rate (see (3.3)) decreases in w, therefore for any value of p the principal can maximize her expected profit rate by lowering w yet safeguarding agent's participation by setting the agent's expected utility rate equal to his reservation utility rate. Although the principal cannot contract directly on the agent's service capacity, she anticipates the agent optimizing his expected utility rate when offered a contract. That is, for any w and p values proposed by the principal, the agent computes the μ that maximizes his expected utility rate and subsequently decides whether to accept the contract or not, by solving the following optimization problem:

$$\max_{\mu\geq 0} u(\mu) = \max_{\mu\geq 0} \left\{ w - \frac{p\lambda}{\lambda+\mu} - \mu + \overline{\eta} p \left(\frac{\lambda}{\lambda+\mu} - \frac{1}{2} \right)_+ \right\} \tag{5.4}$$

The agent's optimal service capacity is denoted by $\mu^*(w,p) = \operatorname{argmax}_{\mu\geq 0} u(\mu)$.

Before proceeding to derive the agent's optimal strategy we introduce some notation:

$$\overline{p}_1 \equiv \frac{\lambda}{1-\overline{\eta}}, \quad \overline{p}_2 \equiv \frac{16\left(2-\overline{\eta}-2\sqrt{1-\overline{\eta}}\right)\lambda}{\overline{\eta}^2} \tag{5.5}$$

and the following identity is verified using the definition of $\overline{p}_2$:

$$\overline{w}_2 \equiv \frac{\overline{\eta}\overline{p}_2}{2} + 2\sqrt{(1-\overline{\eta})\overline{p}_2\lambda} - \lambda = 2\sqrt{\overline{p}_2\lambda} - \lambda \tag{5.6}$$

Note that $\overline{p}_1, \overline{p}_2$ and $\overline{w}_2$ are functions of λ and $\overline{\eta}$. However we suppress $(\lambda, \overline{\eta})$.

Table 5.1 Indicators of the monotonicity and the concavity/convexity of function $u(\mu)$ in (5.3)

Case		$u_\mu(0^+)$	Over $[0,\lambda)$ $u(\mu)$ is	$u_\mu(\lambda^-)$	$u_\mu(\lambda^+)$	Over $(\lambda,+\infty)$ $u(\mu)$ is	$u_\mu(+\infty)$
$\overline{\eta} \in \left(0, \frac{3}{4}\right]$	$p \in \left(0, \frac{\lambda}{1-\overline{\eta}}\right]$	≤ 0	Concave	< 0	< 0	Concave	< 0
	$p \in \left(\frac{\lambda}{1-\overline{\eta}}, 4\lambda\right]^{a}$	> 0	Concave	< 0	≤ 0	Concave	< 0
	$p \in \left(4\lambda, \frac{4\lambda}{1-\overline{\eta}}\right]^{a}$	> 0	Concave	≤ 0	> 0	Concave	< 0
	$p \in \left(\frac{4\lambda}{1-\overline{\eta}}, +\infty\right)$	> 0	Concave	> 0	> 0	Concave	< 0
$\overline{\eta} \in \left(\frac{3}{4}, 1\right)$	$p \in (0, 4\lambda]$	< 0	Concave	< 0	≤ 0	Concave	< 0
	$p \in \left(4\lambda, \frac{\lambda}{1-\overline{\eta}}\right]^{b}$	≤ 0	Concave	< 0	> 0	Concave	< 0
	$p \in \left(\frac{\lambda}{1-\overline{\eta}}, \frac{4\lambda}{1-\overline{\eta}}\right]^{b}$	> 0	Concave	≤ 0	> 0	Concave	< 0
	$p \in \left(\frac{4\lambda}{1-\overline{\eta}}, +\infty\right)$	> 0	Concave	> 0	> 0	Concave	< 0
$\overline{\eta} \in [1, +\infty)$	$p \in (0, 4\lambda]^{c}$	< 0	Convex	< 0	≤ 0	Concave	< 0
	$p \in (4\lambda, +\infty)$	< 0	Convex	< 0	> 0	Concave	< 0

[a]Note that $\overline{\eta} \in (0, 3/4] \Rightarrow 4\lambda/(1-\overline{\eta}) > 4\lambda \geq \lambda/(1-\overline{\eta})$
[b]Note that $\overline{\eta} \in (3/4, 1) \Rightarrow 4\lambda/(1-\overline{\eta}) > \lambda/(1-\overline{\eta}) > 4\lambda$
[c]Note that $\overline{\eta} > 1 \Rightarrow 4\lambda > 0 > \lambda/(1-\overline{\eta}) > 4\lambda/(1-\overline{\eta})$

Next we introduce a number of technical lemmas (see proofs in the Appendix).

Lemma 5.2. *Let* $1 > \overline{\eta} > 0$ *and* $\lambda > 0$.

(a) *If* $\frac{4\left(\sqrt{1-\overline{\eta}/2}-\sqrt{1-\overline{\eta}}\right)^2\lambda}{\overline{\eta}^2} > p > 0$, *then* $0 > \frac{\overline{\eta}p}{2} + 2\sqrt{(1-\overline{\eta})p\lambda} - \lambda$.

(b) *If* $p > \frac{4\left(\sqrt{1-\overline{\eta}/2}-\sqrt{1-\overline{\eta}}\right)^2\lambda}{\overline{\eta}^2}$, *then* $\frac{\overline{\eta}p}{2} + 2\sqrt{(1-\overline{\eta})p\lambda} - \lambda > 0$.

(c) *If* $p = \frac{4\left(\sqrt{1-\overline{\eta}/2}-\sqrt{1-\overline{\eta}}\right)^2\lambda}{\overline{\eta}^2}$, *then* $\frac{\overline{\eta}p}{2} + 2\sqrt{(1-\overline{\eta})p\lambda} - \lambda = 0$.

Lemma 5.3. *Let* $1 > \overline{\eta} > 0$ *and* $\lambda > 0$, *then* $\lambda/(1-\overline{\eta}) > 4\left(\sqrt{1-\overline{\eta}/2}-\sqrt{1-\overline{\eta}}\right)^2 \lambda/\overline{\eta}^2$.

Lemma 5.4. *Let* $2 > \overline{\eta} > 0$ *and* $\lambda > 0$.

(a) *If* $\frac{2\lambda}{2+\overline{\eta}-2\sqrt{2\overline{\eta}}} > p > \frac{2\lambda}{2+\overline{\eta}+2\sqrt{2\overline{\eta}}}$, *then* $0 > \left(1-\frac{\overline{\eta}}{2}\right)p - 2\sqrt{p\lambda} + \lambda$.

(b) *If* $\frac{2\lambda}{2+\overline{\eta}+2\sqrt{2\overline{\eta}}} > p > 0$ *or* $p > \frac{2\lambda}{2+\overline{\eta}-2\sqrt{2\overline{\eta}}}$, *then* $\left(1-\frac{\overline{\eta}}{2}\right)p - 2\sqrt{p\lambda} + \lambda > 0$.

(c) *If* $p = \frac{2\lambda}{2+\overline{\eta}+2\sqrt{2\overline{\eta}}}$ *or* $p = \frac{2\lambda}{2+\overline{\eta}-2\sqrt{2\overline{\eta}}}$, *then* $\left(1-\frac{\overline{\eta}}{2}\right)p - 2\sqrt{p\lambda} + \lambda = 0$.

Lemma 5.5. *Let* $\overline{\eta} > 0$ *and* $\lambda > 0$, *then* $4\lambda > 2\lambda/\left(2+\overline{\eta}+2\sqrt{2\overline{\eta}}\right)$.

Lemma 5.6. *Let* $\lambda > 0$.

(a) *If* $\frac{8}{9} > \overline{\eta} > 0$, *then* $\frac{2\lambda}{2+\overline{\eta}-2\sqrt{2\overline{\eta}}} > \frac{\lambda}{1-\overline{\eta}}$.

(b) *If* $1 > \overline{\eta} > \frac{8}{9}$, *then* $\frac{\lambda}{1-\overline{\eta}} > \frac{2\lambda}{2+\overline{\eta}-2\sqrt{2\overline{\eta}}}$.

(c) *If* $\overline{\eta} = \frac{8}{9}$, *then* $\frac{2\lambda}{2+\overline{\eta}-2\sqrt{2\overline{\eta}}} = \frac{\lambda}{1-\overline{\eta}}$.

Lemma 5.7. *Let* $1 > \overline{\eta} > 0$ *and* $\lambda > 0$.

(a) *If* $\frac{16\left(2-\overline{\eta}-2\sqrt{1-\overline{\eta}}\right)\lambda}{\overline{\eta}^2} > p > 0$, *then* $0 > \frac{\overline{\eta}p}{2} - 2\left(1-\sqrt{1-\overline{\eta}}\right)\sqrt{p\lambda}$.

(b) *If* $p > \frac{16\left(2-\overline{\eta}-2\sqrt{1-\overline{\eta}}\right)\lambda}{\overline{\eta}^2}$, *then* $\frac{\overline{\eta}p}{2} - 2\left(1-\sqrt{1-\overline{\eta}}\right)\sqrt{p\lambda} > 0$.

(c) *If* $p = 0$ *or* $p = \frac{16\left(2-\overline{\eta}-2\sqrt{1-\overline{\eta}}\right)\lambda}{\overline{\eta}^2}$, *then* $\frac{\overline{\eta}p}{2} - 2\left(1-\sqrt{1-\overline{\eta}}\right)\sqrt{p\lambda} = 0$.

Lemma 5.8. *Let* $\lambda > 0$.

(a) *If* $\frac{8}{9} > \overline{\eta} > 0$, *then* $\frac{16\left(2-\overline{\eta}-2\sqrt{1-\overline{\eta}}\right)\lambda}{\overline{\eta}^2} > \frac{\lambda}{1-\overline{\eta}}$.

(b) *If* $1 > \overline{\eta} > \frac{8}{9}$, *then* $\frac{\lambda}{1-\overline{\eta}} > \frac{16\left(2-\overline{\eta}-2\sqrt{1-\overline{\eta}}\right)\lambda}{\overline{\eta}^2}$.

(c) *If* $\overline{\eta} = \frac{8}{9}$, *then* $\frac{16\left(2-\overline{\eta}-2\sqrt{1-\overline{\eta}}\right)\lambda}{\overline{\eta}^2} = \frac{\lambda}{1-\overline{\eta}}$.

Lemma 5.9. *Let* $1 > \overline{\eta} > 0$ *and* $\lambda > 0$, *then* $4\lambda/(1-\overline{\eta}) > 16\left(2-\overline{\eta}-2\sqrt{1-\overline{\eta}}\right)$ $\lambda/\overline{\eta}^2 > 4\lambda$.

Lemmas 5.8 and 5.9 imply $\overline{\eta} \in (0, 3/4) \Rightarrow 4\overline{p}_1 > \overline{p}_2 > 4\lambda \geq \overline{p}_1 > 0$ and $\overline{\eta} \in (3/4, 8/9) \Rightarrow 4\overline{p}_1 > \overline{p}_2 > \overline{p}_1 > 4\lambda > 0$.

We present weakly risk-seeking agent's optimal response to any contract offers $(w, p) \in \mathbb{R}^2_+$ in Proposition 5.10.

Proposition 5.10. *Consider a weakly risk-seeking agent* ($\overline{\eta} \in (0, 8/9)$).

(a) *Given*

$$p \in (0, \overline{p}_1] \text{ and } w \geq \left(1 - \frac{\overline{\eta}}{2}\right) p \tag{5.7}$$

then the agent accepts the contract and installs $\mu^*(w,p) = 0$ *with resulting expected utility rate* $u_A(\mu^*(w,p); w,p) = w - (1 - \overline{\eta}/2)p \geq 0$. *The agent rejects the contract if both* $p \in (0, \overline{p}_1]$ *and* $w \in (0, (1 - \overline{\eta}/2)p)$.

(b) *Given*

$$p \in (\overline{p}_1, \overline{p}_2) \text{ and } w \geq \frac{\overline{\eta}p}{2} + 2\sqrt{(1-\overline{\eta})p\lambda} - \lambda \tag{5.8}$$

then the agent accepts the contract and installs $\mu^*(w,p) = \sqrt{(1-\overline{\eta})p\lambda} - \lambda$ *with resulting expected utility rate* $u_A(\mu^*(w,p); w,p) = w - \overline{\eta}p/2 - 2\sqrt{(1-\overline{\eta})p\lambda} + \lambda \geq 0$. *The agent rejects the contract if both* $p \in (\overline{p}_1, \overline{p}_2)$ *and* $w \in \left(0, \overline{\eta}p/2 + 2\sqrt{(1-\overline{\eta})p\lambda} - \lambda\right)$.

(c) *Given*

$$p = \overline{p}_2 \text{ and } w \geq \overline{w}_2 \tag{5.9}$$

then the agent accepts the contract and is indifferent about installing either $\mu^*(w,p) = \sqrt{(1-\overline{\eta})\overline{p}_2\lambda} - \lambda$ *or* $\mu^*(w,p) = \sqrt{\overline{p}_2\lambda} - \lambda$. *In both cases the agent receives* $u_A(\mu^*(w,p); w,p) = w - \overline{w}_2 \geq 0$. *If* $r \in (0, \overline{p}_2)$, *then there exists* w^* *such that* $\left((w^*, \overline{p}_2), \mu^* = \sqrt{(1-\overline{\eta})\overline{p}_2\lambda} - \lambda\right)$ *is the unique admissible solution (see Definition 5.1). If* $r = \overline{p}_2$, *there exists* w^* *such* *that* $\left((w^*, \overline{p}_2), \mu^* = \sqrt{\overline{p}_2\lambda} - \lambda\right)$ *and* $\left((w^*, \overline{p}_2), \mu^* = \sqrt{(1-\overline{\eta})\overline{p}_2\lambda} - \lambda\right)$ *are both admissible solutions. If* $r > \overline{p}_2$, *then there exists* w^* *such that* $\left((w^*, \overline{p}_2), \mu^* = \sqrt{\overline{p}_2\lambda} - \lambda\right)$ *is the unique admissible solution (for proof see Proposition 5.13). The agent rejects the contract if both* $p = \overline{p}_2$ *and* $w \in (0, \overline{w}_2)$.

(d) *Given*

$$p > \overline{p}_2 \text{ and } w \geq 2\sqrt{p\lambda} - \lambda \tag{5.10}$$

then the agent accepts the contract and installs $\mu^*(w,p) = \sqrt{p\lambda} - \lambda$ *with resulting expected utility rate* $u_A(\mu^*(w,p); w, p) = w - 2\sqrt{p\lambda} + \lambda \geq 0$. *The agent rejects the contract if both* $p > \overline{p}_2$ *and* $w \in \left(0, 2\sqrt{p\lambda} - \lambda\right)$.

Proof. According to Table 5.1, the behavior of $u(\mu)$ when $\overline{\eta} \in (0, 3/4]$ versus $\overline{\eta} \in (3/4, 8/9)$ is different. Therefore we prove the proposition separately for $\overline{\eta} \in (0, 3/4]$ and $\overline{\eta} \in (3/4, 8/9)$.

Case $\overline{\eta} \in (0, 3/4]$: According to Lemmas 5.8 part (a) and 5.9, $4\overline{p}_1 > \overline{p}_2 > 4\lambda \geq \overline{p}_1 > 0$. Figure 5.2 depicts the shape of $u(\mu)$ when $\overline{\eta} \in (0, 3/4]$ and the value of p falls in different ranges. The structure of the proof when $\overline{\eta} \in (0, 3/4]$ is depicted in Fig. 5.3.

Case $p \in (0, \overline{p}_1]$: According to Table 5.1, $u(\mu)$ is decreasing with respect to $\mu \geq 0$. Thus the agent's optimal service capacity is $\mu^*(w,p) = 0$ and from (5.3) $u(\mu^*(w,p)) = w - (1 - \overline{\eta}/2)p$. Note that $1 - \overline{\eta}/2 > 0$.

Subcase $w \in (0, (1 - \overline{\eta}/2)p)$: $u(\mu^*(w,p)) < 0$, therefore the agent rejects the contract.

Subcase $w \geq (1 - \overline{\eta}/2)p$: $u(\mu^*(w,p)) \geq 0$, thus the agent would accept the contract if offered.

Case $p \in (\overline{p}_1, 4\lambda]$: According to Table 5.1, the service capacity that maximizes $u(\mu)$ lies in $(0, \lambda)$. μ^* is computed from first order condition $du(\mu)/d\mu|_{\mu=\mu^*(w,p)} = 0 \Rightarrow \mu^*(w,p) = \sqrt{(1-\overline{\eta})p\lambda} - \lambda > 0$ and from (5.3) $u(\mu^*(w,p)) = w - \overline{\eta}p/2 - 2\sqrt{(1-\overline{\eta})p\lambda} + \lambda$. According to Lemmas 5.2 part (b) and 5.3, $\overline{\eta}p/2 + 2\sqrt{(1-\overline{\eta})p\lambda} - \lambda > 0$, therefore we examine the following subcases.

Subcase $w \in \left(0, \overline{\eta}p/2 + 2\sqrt{(1-\overline{\eta})p\lambda} - \lambda\right)$: $u(\mu^*(w,p)) < 0$, therefore the agent rejects the contract.

Subcase $w \geq \overline{\eta}p/2 + 2\sqrt{(1-\overline{\eta})p\lambda} - \lambda$: $u(\mu^*(w,p)) \geq 0$, thus the agent would accept the contract if offered.

Case $p \in (4\lambda, 4\overline{p}_1]$: According to Table 5.1, there is a service capacity that maximizes $u(\mu)$ for $\mu \in (0, \lambda]$ and a service capacity that maximizes $u(\mu)$ for $\mu > \lambda$. Denote the optimal service capacity in $(0, \lambda]$ by $\mu^*_{(0,\lambda]}(w,p)$. From the first order condition the optimal service capacity is $\mu^*_{(0,\lambda]}(w,p) = \sqrt{(1-\overline{\eta})p\lambda} - \lambda$

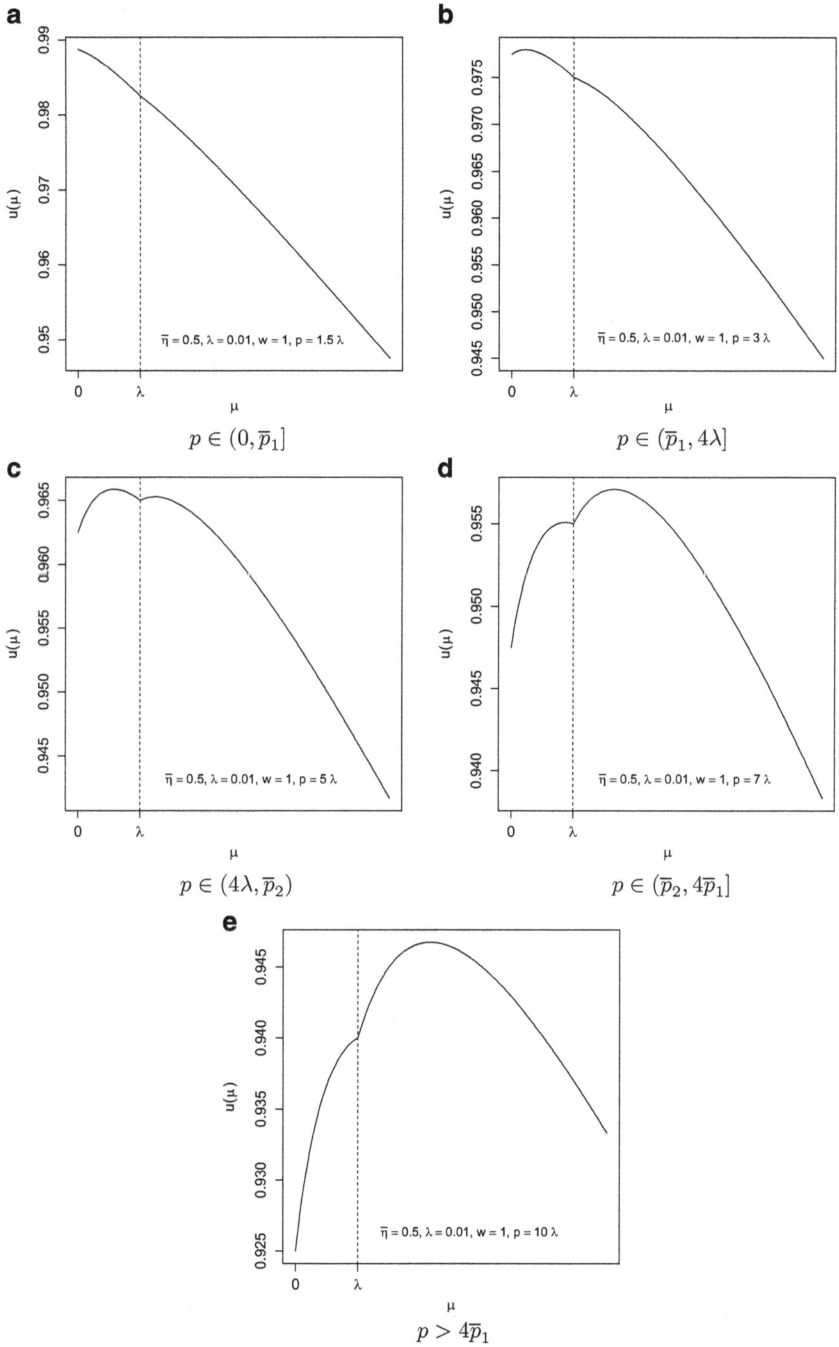

Fig. 5.2 Illustration of the forms of $u(\mu)$ when $\overline{\eta} \in (0, 3/4]$

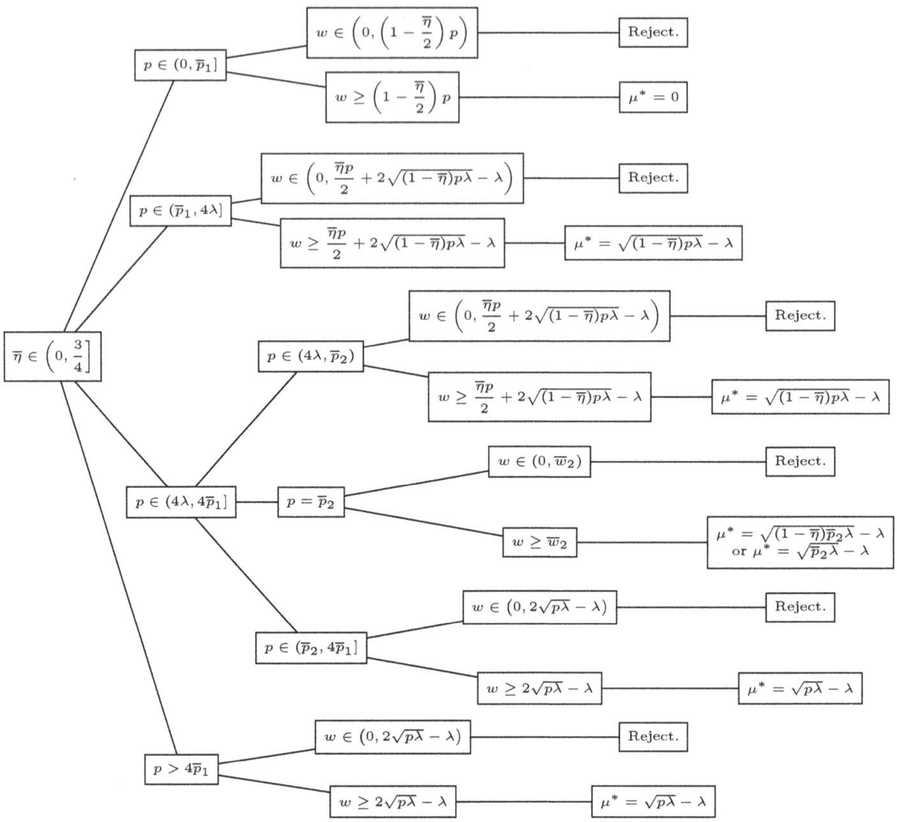

Fig. 5.3 Structure of the proof for Proposition 5.10 when $\overline{\eta} \in (0, 3/4]$

and from Eq. (5.3) $u\left(\mu^*_{(0,\lambda]}(w,p)\right) = w - \overline{\eta}p/2 - 2\sqrt{(1-\overline{\eta})p\lambda} + \lambda$. Denote the optimal service capacity for $\mu > \lambda$ by $\mu^*_{\lambda}(w,p)$, which is obtained from first order condition $du(\mu)/d\mu|_{\mu=\mu^*_{\lambda}(w,p)} = 0 \Rightarrow \mu^*_{\lambda}(w,p) = \sqrt{p\lambda} - \lambda$ and from Eq. (5.3) $u\left(\mu^*_{\lambda}(w,p)\right) = w - 2\sqrt{p\lambda} + \lambda$. The agent has a choice of two service capacities and he installs the one that generates a higher expected utility rate. Note that $u\left(\mu^*_{\lambda}(w,p)\right) - u\left(\mu^*_{(0,\lambda]}(w,p)\right) = \overline{\eta}p/2 - 2\left(1 - \sqrt{1-\overline{\eta}}\right)\sqrt{p\lambda}$. According to Lemma 5.9, $4\overline{p}_1 > \overline{p}_2 > 4\lambda$, therefore we examine the following subcases.

Subcase $p \in (4\lambda, \overline{p}_2)$: By Lemma 5.7 part (a), $u\left(\mu^*_{(0,\lambda]}(w,p)\right) > u\left(\mu^*_{\lambda}(w,p)\right)$, therefore the agent's optimal service capacity is $\mu^*(w,p) = \sqrt{(1-\overline{\eta})p\lambda} - \lambda$ and $u(\mu^*(w,p)) = w - \overline{\eta}p/2 - 2\sqrt{(1-\overline{\eta})p\lambda} + \lambda$. According to Lemmas 5.2 part (a) and 5.3, $\overline{\eta}p/2 + 2\sqrt{(1-\overline{\eta})p\lambda} - \lambda > 0$, therefore we examine the following subcases.

Subsubcase $w \in \left(0, \overline{\eta}p/2 + 2\sqrt{(1-\overline{\eta})p\lambda} - \lambda\right)$: $u(\mu^*(w,p)) < 0$, thus the agent rejects the contract.

Subsubcase $w \geq \overline{\eta}p/2 + 2\sqrt{(1-\overline{\eta})p\lambda} - \lambda$: $u(\mu^*(w,p)) \geq 0$, therefore the agent would accept the contract if offered.

Subcase $p = \overline{p}_2$: According to Lemma 5.7 part (c), $u\left(\mu^*_{(0,\lambda]}(w,p)\right) = u\big(\mu^*_\lambda(w,p)\big)$, indicating that installing either $\mu^*_{(0,\lambda]}(w,\overline{p}_2)$ or $\mu^*_\lambda(w,\overline{p}_2)$ leads to the same agent's expected utility rate. Therefore the agent is indifferent about installing $\mu^*(w,p) = \sqrt{(1-\overline{\eta})p\lambda} - \lambda$ or $\mu^*(w,p) = \sqrt{p\lambda} - \lambda$. Still, the capacity value leads to admissible solutions (see Proposition 5.13). Recall the definition of $\overline{w}_2$ from (5.6). By Lemma 5.2, $\overline{p}_2 > 4\lambda > \overline{p}_1 \Rightarrow \overline{w}_2 = \overline{\eta}\overline{p}_2/2 + 2\sqrt{(1-\overline{\eta})\overline{p}_2\lambda} - \lambda > 0$, therefore we examine the following subcases.

Subsubcase $w \in (0, \overline{w}_2)$: $u(\mu^*(w,p)) < 0$, therefore the agent rejects the contract.

Subsubcase $w \geq \overline{w}_2$: $u(\mu^*(w,p)) \geq 0$, therefore the agent would accept the contract if offered.

Subcase $p \in (\overline{p}_2, 4\overline{p}_1]$: From Lemma 5.7 part (b), $u\left(\mu^*_\lambda(w,p)\right) > u\big(\mu^*_{(0,\lambda]}(w,p)\big)$, therefore the agent's optimal service capacity is $\mu^*(w,p) = \sqrt{p\lambda} - \lambda$ and $u(\mu^*(w,p)) = w - 2\sqrt{p\lambda} + \lambda$. Since $p > \overline{p}_2 > 4\lambda \Rightarrow 2\sqrt{p\lambda} - \lambda > 3\lambda > 0$, therefore we examine the following subcases.

Subsubcase $w \in \left(0, 2\sqrt{p\lambda} - \lambda\right)$: $u(\mu^*(w,p)) < 0$, thus the agent rejects the contract.

Subsubcase $w \geq 2\sqrt{p\lambda} - \lambda$: $u(\mu^*(w,p)) \geq 0$, therefore the agent would accept the contract if offered.

Case $p > 4\overline{p}_1$: According to Table 5.1, the service capacity that maximizes $u(\mu)$ satisfies $\mu > \lambda$. From the first order condition the agent's optimal service capacity is $\mu^*(w,p) = \sqrt{p\lambda} - \lambda$ and from Eq. (5.3) $u(\mu^*(w,p)) = w - 2\sqrt{p\lambda} + \lambda$. Since $p > 4\overline{p}_1 > 4\lambda$, therefore $2\sqrt{p\lambda} - \lambda > 3\lambda > 0$ and we examine the following subcases.

Subcase $w \in \left(0, 2\sqrt{p\lambda} - \lambda\right)$: $u(\mu^*(w,p)) < 0$, thus the agent rejects the contract.

Subcase $w \geq 2\sqrt{p\lambda} - \lambda$: $u(\mu^*(w,p)) \geq 0$, therefore the agent would accept the contract if offered.

This completes the proof for Proposition 5.10 when $\overline{\eta} \in (0, 3/4]$.

Case $\overline{\eta} \in (3/4, 8/9)$: According to Lemmas 5.8 and 5.9, $4\overline{p}_1 > \overline{p}_2 > \overline{p}_1 > 4\lambda > 0$. Figure 5.4 depicts the shape of $u(\mu)$ when $\overline{\eta} \in (3/4, 8/9)$ and the value of p falls in different ranges. The structure of the proof when $\overline{\eta} \in (3/4, 8/9)$ is depicted in Fig. 5.5.

Case $p \in (0, 4\lambda]$: According to Table 5.1, $u(\mu)$ is decreasing with respect to $\mu \geq 0$. Therefore the agent's optimal service capacity is $\mu^*(w,p) = 0$ and from Eq. (5.3) $u(\mu^*(w,p)) = w - (1 - \overline{\eta}/2)p$. Note that $1 - \overline{\eta}/2 > 0$.

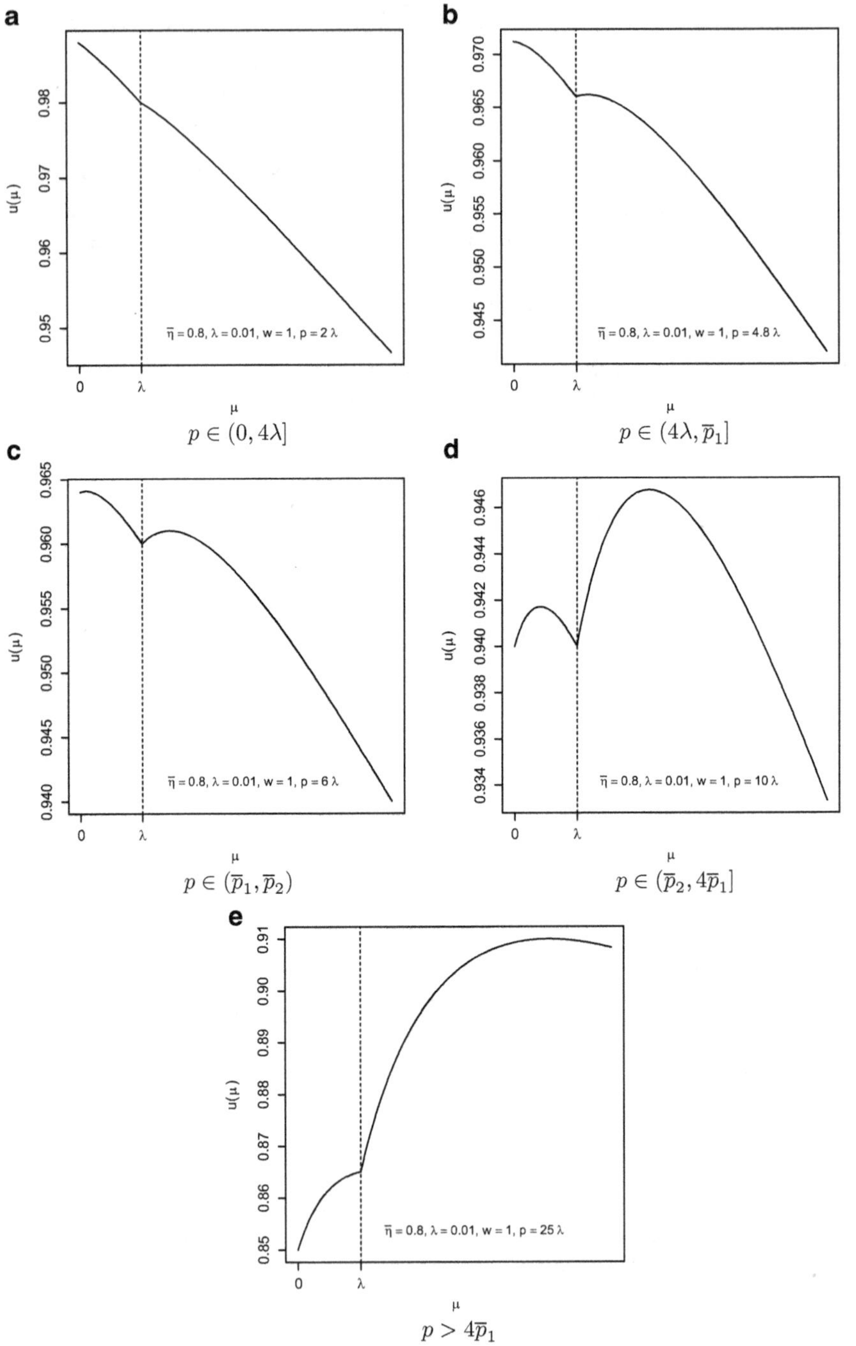

Fig. 5.4 Illustration of the forms of $u(\mu)$ when $\overline{\eta} \in (3/4, 8/9)$

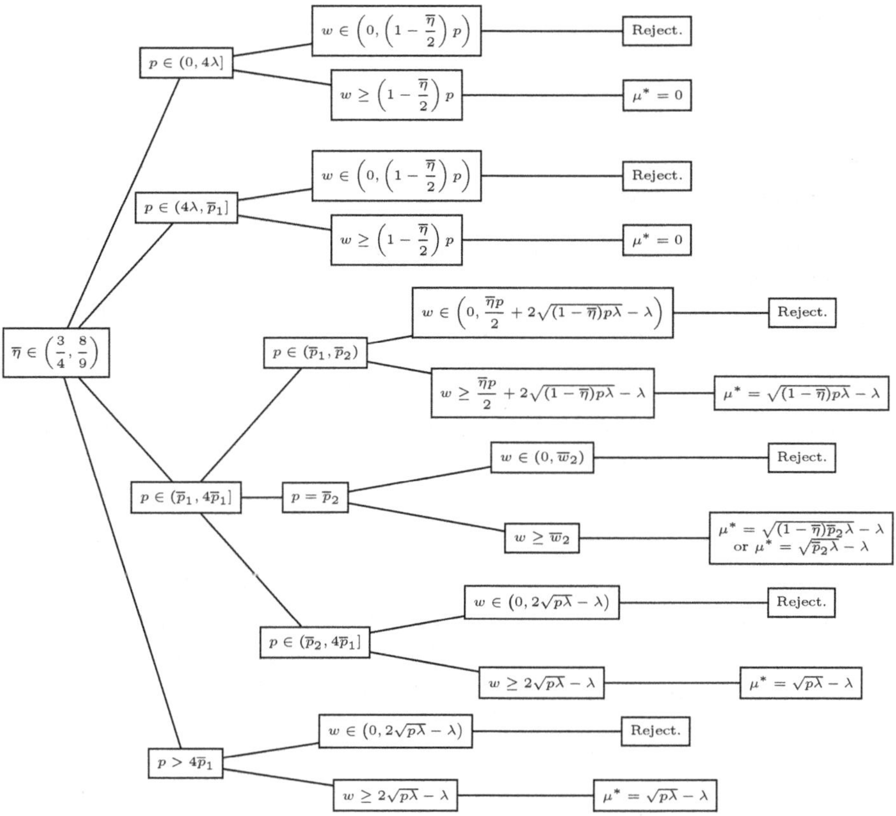

Fig. 5.5 Structure of the proof for Proposition 5.10 when $\overline{\eta} \in (3/4, 8/9)$

Subcase $w \in (0, (1-\overline{\eta}/2)p)$: $u(\mu^*(w,p)) < 0$, therefore the agent rejects the contract.

Subcase $w \geq (1-\overline{\eta}/2)p$: $u(\mu^*(w,p)) \geq 0$, thus the agent would accept the contract if offered.

Case $p \in (4\lambda, \overline{p}_1]$: According to Table 5.1, there is a service capacity that maximizes $u(\mu)$ for $\mu \in [0,\lambda)$ and a service capacity that maximizes $u(\mu)$ for $\mu > \lambda$. Denote the optimal service capacity in $[0,\lambda)$ by $\mu^*_{[0,\lambda)}(w,p)$. Since $u(\mu)$ is decreasing with respect to μ over $[0,\lambda)$, therefore $\mu^*_{[0,\lambda)}(w,p) = 0$ and from (5.3) $u\left(\mu^*_{[0,\lambda)}(w,p)\right) = w - (1-\overline{\eta}/2)p$. Denote the optimal service capacity for $\mu > \lambda$ by $\mu^*_{\lambda}(w,p)$. From first order condition $\mu^*_{\lambda}(w,p) = \sqrt{p\lambda} - \lambda$ and from (5.3) $u\left(\mu^*_{\lambda}(w,p)\right) = w - 2\sqrt{p\lambda} + \lambda$. The agent has to choose one of the two service capacities and he installs the one with higher expected utility rate. Note that $u\left(\mu^*_{\lambda}(w,p)\right) - u\left(\mu^*_{[0,\lambda)}(w,p)\right) = (1-\overline{\eta}/2)p - 2\sqrt{p\lambda} +$

λ. According to Lemma 5.5, $4\lambda > 2\lambda/\left(2+\overline{\eta}+2\sqrt{2\overline{\eta}}\right)$ and according to Lemma 5.6, $2\lambda/\left(2+\overline{\eta}-2\sqrt{2\overline{\eta}}\right) > \overline{p}_1$. Therefore according to Lemma 5.4 part (a), $u\left(\mu^*_{[0,\lambda)}(w,p)\right) > u\left(\mu^*_{\lambda}(w,p)\right)$, the agent's optimal service capacity is $\mu^*(w,p) = 0$ and $u(\mu^*(w,p)) = w-(1-\overline{\eta}/2)p$. Note that $1-\overline{\eta}/2 > 0$, therefore we examine the following subcases.

Subcase $w \in (0,(1-\overline{\eta}/2)p)$: $u(\mu^*(w,p)) < 0$, therefore the agent rejects the contract.

Subcase $w \geq (1-\overline{\eta}/2)p$: $u(\mu^*(w,p)) \geq 0$, thus the agent would accept the contract if offered.

Case $p \in (\overline{p}_1, 4\overline{p}_1]$: According to Table 5.1, there is a service capacity that maximizes $u(\mu)$ for $\mu \in (0,\lambda]$ and a service capacity that maximizes $u(\mu)$ for $\mu > \lambda$. Denote the optimal service capacity in $(0,\lambda]$ by $\mu^*_{(0,\lambda]}(w,p)$. From first order condition the optimal service capacity is $\mu^*_{(0,\lambda]}(w,p) = \sqrt{(1-\overline{\eta})p\lambda} - \lambda$ and from (5.3) $u\left(\mu^*_{(0,\lambda]}(w,p)\right) = w - \overline{\eta}p/2 - 2\sqrt{(1-\overline{\eta})p\lambda} + \lambda$. Denote the optimal service capacity for $\mu > \lambda$ by $\mu^*_{\lambda}(w,p)$. From first order condition $\mu^*_{\lambda}(w,p) = \sqrt{p\lambda} - \lambda$ and from (5.3) $u\left(\mu^*_{\lambda}(w,p)\right) = w - 2\sqrt{p\lambda} + \lambda$. The agent has a choice of two service capacities and he installs the one that generates a higher expected utility rate. Note that $u\left(\mu^*_{\lambda}(w,p)\right) - u\left(\mu^*_{(0,\lambda]}(w,p)\right) = \overline{\eta}p/2 - 2\left(1-\sqrt{1-\overline{\eta}}\right)\sqrt{p\lambda}$. According to Lemmas 5.8 and 5.9, $4\overline{p}_1 > \overline{p}_2 > \overline{p}_1$, therefore we examine the following subcases.

Subcase $p \in (\overline{p}_1, \overline{p}_2)$: By Lemma 5.7 part (a), $u\left(\mu^*_{(0,\lambda]}(w,p)\right) > u\left(\mu^*_{\lambda}(w,p)\right)$, therefore the agent's optimal service capacity is $\mu^*(w,p) = \sqrt{(1-\overline{\eta})p\lambda} - \lambda$ and $u(\mu^*(w,p)) = w - \overline{\eta}p/2 - 2\sqrt{(1-\overline{\eta})p\lambda} + \lambda$. According to Lemmas 5.2 and 5.3, $p > \overline{p}_1 \Rightarrow \overline{\eta}p/2 + 2\sqrt{(1-\overline{\eta})p\lambda} - \lambda > 0$, therefore we examine the following subcases.

Subsubcase $w \in \left(0, \overline{\eta}p/2 + 2\sqrt{(1-\overline{\eta})p\lambda} - \lambda\right)$: $u(\mu^*(w,p)) < 0$, thus the agent rejects the contract.

Subsubcase $w \geq \overline{\eta}p/2 + 2\sqrt{(1-\overline{\eta})p\lambda} - \lambda$: $u(\mu^*(w,p)) \geq 0$, therefore the agent would accept the contract if offered.

Subcase $p = \overline{p}_2$: According to Lemma 5.7 part (c), $u\left(\mu^*_{(0,\lambda]}(w,p)\right) = u\left(\mu^*_{\lambda}(w,p)\right)$, indicating that installing $\mu^*_{(0,\lambda]}(w,\overline{p}_2)$ or $\mu^*_{\lambda}(w,\overline{p}_2)$ leads to the same agent's expected utility rate. Therefore the agent is indifferent about installing $\mu^*(w,p) = \sqrt{(1-\overline{\eta})p\lambda} - \lambda$ or $\mu^*(w,p) = \sqrt{p\lambda} - \lambda$. Still, the capacity value has to lead to admissible solutions (see Proposition 5.13).

Recall the definition of $\overline{w}_2$ in (5.6). According to Lemma 5.2, $\overline{p}_2 > \overline{p}_1 \Rightarrow \overline{w}_2 = \overline{\eta}\overline{p}_2/2 + 2\sqrt{(1-\overline{\eta})\overline{p}_2\lambda} - \lambda > 0$, therefore we examine the following subcases.

Subsubcase $w \in (0, \overline{w}_2)$: $u(\mu^*(w,p)) < 0$, therefore the agent rejects the contract.

Subsubcase $w \geq \overline{w}_2$: $u(\mu^*(w,p)) \geq 0$, so the agent would accept the contract if offered.

Subcase $p \in (\overline{p}_2, 4\overline{p}_1]$: By Lemma 5.7 part (b), $u\left(\mu^*_\lambda(w,p)\right) > u\left(\mu^*_{(0,\lambda]}(w,p)\right)$, therefore the agent's optimal service capacity is $\mu^*(w,p) = \sqrt{p\lambda} - \lambda$ and $u(\mu^*(w,p)) = w - 2\sqrt{p\lambda} + \lambda$. Since $p > \overline{p}_2 > 4\lambda \Rightarrow 2\sqrt{p\lambda} - \lambda > 3\lambda > 0$, therefore we examine the following subcases.

Subsubcase $w \in \left(0, 2\sqrt{p\lambda} - \lambda\right)$: $u(\mu^*(w,p)) < 0$, thus the agent rejects the contract.

Subsubcase $w \geq 2\sqrt{p\lambda} - \lambda$: $u(\mu^*(w,p)) \geq 0$, therefore the agent would accept the contract if offered.

Case $p > 4\overline{p}_1$: According to Table 5.1, the service capacity that maximizes $u(\mu)$ satisfies $\mu > \lambda$. From the first order condition the agent's optimal service capacity is $\mu^*(w,p) = \sqrt{p\lambda} - \lambda$ and from (5.3) $u(\mu^*(w,p)) = w - 2\sqrt{p\lambda} + \lambda$. Since $p > 4\overline{p}_1 > 4\lambda$, therefore $2\sqrt{p\lambda} - \lambda > 3\lambda > 0$ and we examine the following subcases.

Subcase $w \in \left(0, 2\sqrt{p\lambda} - \lambda\right)$: $u(\mu^*(w,p)) < 0$, thus the agent rejects the contract.

Subcase $w \geq 2\sqrt{p\lambda} - \lambda$: $u(\mu^*(w,p)) \geq 0$, therefore the agent would accept the contract if offered.

This complete the proof for Proposition 5.10 when $\overline{\eta} \in (3/4, 8/9)$. □

To summarize: Given exogenous market conditions that enable a mutually beneficial contract between a principal and weakly risk-seeking agent (see Theorem 5.17 later), the agent determines his service capacity by using one of only two formulas:

$$\mu^* = \sqrt{(1-\overline{\eta})p\lambda} - \lambda > 0 \text{ or } \mu^*(w,p) = \sqrt{p\lambda} - \lambda > 0$$

The conditions when a weakly risk-seeking agent accepts the contract can be depicted by the shaded areas in Fig. 5.6, where $\overline{\eta} = 0.5$. The three shaded areas with different grey scales represent conditions (5.7), (5.8) and (5.10) under which the agent accepts the contract but responds differently. The lower bound function of the shaded area (denoted by $w_0(p)$) represents the set of offers of zero expected utility rate for the agent. The $w_0(p)$ line is defined as follows:

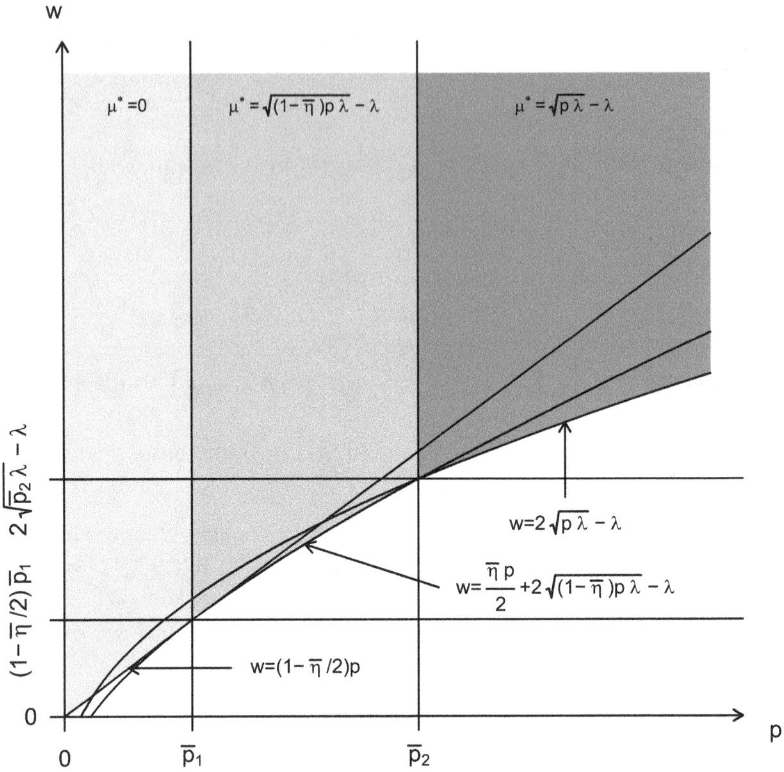

Fig. 5.6 Conditions when a weakly risk-seeking agent accepts the contract with $\overline{\eta} = 0.5$

$$w_0(p) = \begin{cases} \left(1 - \dfrac{\overline{\eta}}{2}\right) p \text{ when } p \in (0, \overline{p}_1] \\ \dfrac{\overline{\eta}p}{2} + 2\sqrt{(1-\overline{\eta})p\lambda} - \lambda \text{ when } p \in (\overline{p}_1, \overline{p}_2] \\ 2\sqrt{p\lambda} - \lambda \text{ when } p > \overline{p}_2 \end{cases}$$

Since $\lim_{p \to \overline{p}_1^-} w_0(p) = \lim_{p \to \overline{p}_1^+} w_0(p) = (1 - \overline{\eta}/2)\overline{p}_1$ and $\lim_{p \to \overline{p}_2^-} w_0(p) = \lim_{p \to \overline{p}_2^+} w_0(p) = \overline{\eta}\overline{p}_2/2 + 2\sqrt{(1-\overline{\eta})\overline{p}_2\lambda} - \lambda$, therefore $w_0(p)$ is continuous everywhere over interval $p \in \mathbb{R}_+$. Since $\lim_{p \to \overline{p}_1^-} dw_0(p)/dp = \lim_{p \to \overline{p}_1^+} dw_0(p)/dp = 1 - \overline{\eta}/2$, therefore $w_0(p)$ is differentiable at $p = \overline{p}_1$. However since $\lim_{p \to \overline{p}_2^-} dw_0(p)/dp = \overline{\eta}\left(2 - \sqrt{1-\overline{\eta}}\right)/4\left(1 - \sqrt{1-\overline{\eta}}\right) \neq \overline{\eta}/4\left(1 - \sqrt{1-\overline{\eta}}\right) = \lim_{p \to \overline{p}_2^+} dw_0(p)/dp$, therefore $w_0(p)$ is not differentiable at $p = \overline{p}_2$.

5.1.1 Sensitivity Analysis of a Weakly Risk-Seeking Agent's Optimal Strategy

A principal does not propose a contract that will be accepted by the agent but results in zero service capacity. Therefore the only viable cases when the agent accepts the contract and installs positive service capacities are: $\mu^*(w,p) = \sqrt{(1-\overline{\eta})p\lambda} - \lambda$ or $\mu^*(w,p) = \sqrt{p\lambda} - \lambda$.

First the case when a weakly risk-seeking agent installs $\mu^*(w,p) = \sqrt{(1-\overline{\eta})p\lambda} - \lambda$. According to (5.8) the compensation rate w is bounded below by $\overline{\eta}p/2 + 2\sqrt{(1-\overline{\eta})p\lambda} - \lambda = pP(1) - \overline{\eta}p\,(P(1) - 1/2) + \mu^*(w,p)$, with the term $pP(1)$ representing the expected penalty rate charged by the principal and the term $\overline{\eta}p\,(P(1) - 1/2)$ representing the expected risk rate perceived by the agent when the optimal capacity is installed. It dictates that the agent be reimbursed for the expected penalty rate and the cost of the optimal service capacity discounted by his perceived risk rate in exchange.

The optimal service capacity $\sqrt{(1-\overline{\eta})p\lambda} - \lambda$ depends on p, λ, and $\overline{\eta}$. Its derivatives are:

$$\frac{\partial \mu^*}{\partial p} = \sqrt{\frac{(1-\overline{\eta})\lambda}{4p}} > 0,\ \frac{\partial \mu^*}{\partial \lambda} = \sqrt{\frac{(1-\overline{\eta})p}{4\lambda}} - 1 \text{ and } \frac{\partial \mu^*}{\partial \overline{\eta}} = -\sqrt{\frac{p\lambda}{4(1-\overline{\eta})}} < 0$$

The above derivatives indicate that given a λ and η the agent will increase the service capacity when the penalty rate increases. Note that $\sqrt{(1-\overline{\eta})p\lambda} - \lambda$, as a function of λ, decreases when $\lambda > (1-\overline{\eta})p/4$. From conditions (5.8) and (5.9) the agent installs service capacity $\sqrt{(1-\overline{\eta})p\lambda} - \lambda$ when $p \in (\overline{p}_1, \overline{p}_2]$, and according to Lemma 5.9 we have $4\overline{p}_1 > \overline{p}_2$. Therefore we have $4\lambda/(1-\overline{\eta}) = 4\overline{p}_1 > p \Rightarrow \lambda > (1-\overline{\eta})p/4 \Rightarrow 0 > \partial\mu^*/\partial\lambda$. Thus, given the penalty rate and the risk coefficient, the agent will decrease the service capacity when the failure rate increases. Given a penalty rate and a failure rate, the agent will reduce the service capacity when he is more risk-seeking.

The agent's optimal expected utility rate when installing capacity $\mu^*(w,p) = \sqrt{(1-\overline{\eta})p\lambda} - \lambda$ is $u_A^* \equiv u_A(\mu^*(w,p); w,p) = w - \overline{\eta}p/2 - 2\sqrt{(1-\overline{\eta})p\lambda} + \lambda$, and it depends on w, p, $\overline{\eta}$ and λ. Note that $\partial u_A^*/\partial w = -1 < 0$, $\partial u_A^*/\partial p = -\overline{\eta}/2 - \sqrt{(1-\overline{\eta})\lambda/p} < 0$, indicating that the agent's optimal expected utility rate decreases with the compensation rate and the penalty rate. Note that $\partial u_A^*/\partial\overline{\eta} = -\sqrt{p}\left(\sqrt{p} - \sqrt{4\overline{p}_1}\right)/2$. From Proposition 5.10 $p < \overline{p}_2 < 4\overline{p}_1 \Rightarrow \sqrt{p} < \sqrt{4\overline{p}_1}$, therefore the agent's optimal expected utility rate increases with his risk intensity. Note that $\partial u_A^*/\partial\lambda = -\left(\sqrt{p} - \sqrt{\overline{p}_1}\right)/\sqrt{\overline{p}_1}$, and from Proposition 4.23 $p > \overline{p}_1 \Rightarrow \sqrt{p} - \sqrt{\overline{p}_1} > 0$, therefore the agent's optimal expected utility rate decreases with the failure rate.

Then the case when a weakly risk-seeking agent installs $\mu^*(w,p) = \sqrt{p\lambda} - \lambda$. In this case the agent's optimal strategy is identical to the optimal strategy when he is risk-neutral. According to (5.10) the w is bounded below by $2\sqrt{p\lambda} - \lambda = pP(1) +$

$\mu^*(w,p)$, with the term $pP(1)$ representing the expected penalty rate charged by the principal. It indicates that the agent will have to be reimbursed for the expected penalty rate and the cost of the optimal service capacity.

The optimal service capacity $\sqrt{p\lambda} - \lambda$ depends on the penalty rate p and the failure rate λ. Its derivatives are $\partial\mu^*/\partial p = \sqrt{\lambda/4p} > 0$ and $\partial\mu^*/\partial\lambda = \sqrt{p/4\lambda} - 1$. These derivatives imply that given λ, the agent will increase the service capacity when the penalty rate increases. Note that $\sqrt{p\lambda} - \lambda$, as a function of λ, increases when $p/4 > \lambda$. From conditions (5.9) and (5.10) the agent installs service capacity $\sqrt{p\lambda} - \lambda$ when $p \geq \overline{p}_2$, and according to Lemma 5.9 we have $\overline{p}_2 > 4\lambda$. Therefore we have $p > 4\lambda \Rightarrow p/4 > \lambda \Rightarrow \partial\mu^*/\partial\lambda > 0$. Thus, given p, an agent will increase μ when λ increases.

The agent's optimal expected utility rate when installing capacity $\mu^*(w,p) = \sqrt{p\lambda} - \lambda$ is $u_A^* \equiv u_A(\mu^*(w,p); w, p) = w - 2\sqrt{p\lambda} + \lambda$, and it depends on w, p and λ only. Note that $\partial u_A^*/\partial w = -1 < 0$, $\partial u_A^*/\partial p = -\sqrt{\lambda/p} < 0$, indicating that the agent's optimal expected utility rate decreases with the compensation rate and the penalty rate. Note that $\partial u_A^*/\partial\lambda = -\sqrt{p/\lambda} + 1$, and from Proposition 5.10 $p \geq \overline{p}_2 > 4\lambda \Rightarrow -\sqrt{p/\lambda} + 1 < 0$, therefore the agent's optimal expected utility rate also decreases with the failure rate.

Summary: Recall that given the set of contract offers $\{(w,p) : p \in (0,\lambda], w \geq p\}$ a risk-neutral agent would accept the contract, install $\mu^*(w,p) = 0$ and receive expected utility rate $u(\mu^*(w,p); w,p) = w - p$. Given the set of offers $\left\{(w,p) : p > \lambda, w \geq 2\sqrt{p\lambda} - \lambda\right\}$ he would accept the contract, install $\mu^*(w,p) = \sqrt{p\lambda} - \lambda$ and receive expected utility rate $u(\mu^*(w,p); w,p) = w - 2\sqrt{p\lambda} + \lambda$. By comparing the optimal capacities of a weakly risk-seeking agent to that of a risk-neutral agent, three conclusions are drawn.

1. The principal has to set a higher penalty rate p in order to induce a weakly risk-seeking agent to install a positive service capacity versus a risk-neutral agent ($p > \lambda$ for risk-neutral agent, $p > \lambda/(1-\overline{\eta})$ for weakly risk-seeking agent).
2. When p is relatively low, μ plays a more prominent role in the utility of a weakly risk-seeking agent who therefore installs a μ lower than that when he is risk-neutral ($\sqrt{p\lambda} - \lambda > \sqrt{(1-\overline{\eta})p\lambda} - \lambda$). As p increases, the weakly risk-seeking agent installs μ that is identical to the one for risk-neutral agent ($\sqrt{p\lambda} - \lambda$).
3. Weakly risk-seeking agent is not worse off.

This conclusion is restated in Proposition 5.11.

Proposition 5.11. *Given w and p, an agent who accepts the contract and installs a positive service capacity has a non-decreasing expected utility rate with $\overline{\eta}$ for $\overline{\eta} \in [0, 8/9)$.*

Proof. Recall that when the compensation rate w and the penalty rate p satisfy conditions (5.8) and (5.9), the agent installs service capacity $\mu^*(w,p) = \sqrt{(1-\overline{\eta})p\lambda} - \lambda > 0$, and the agent's expected utility rate is $u(\mu^*(w,p)) = w - \overline{\eta}p/2 - 2\sqrt{(1-\overline{\eta})p\lambda} + \lambda$. Note that $\partial u/\partial\overline{\eta} = -p/2 + \sqrt{p\lambda/(1-\overline{\eta})} =$

$-\sqrt{p}\left(\sqrt{p}-\sqrt{4\overline{p}_1}\right)/2$. According to Lemma 5.9, $4\overline{p}_1 > \overline{p}_2 \geq p$, therefore $\partial u/\partial\overline{\eta} > 0$. When the compensation rate w and the penalty rate p satisfy conditions (5.9) and (5.10), the agent installs service capacity $\mu^*(w,p) = \sqrt{p\lambda} - \lambda > 0$, and the agent's expected utility rate is $u(\mu^*(w,p)) = w - 2\sqrt{p\lambda} + \lambda$, therefore $\partial u/\partial\overline{\eta} = 0$. □

Corollary 5.12. *Given w and p, an agent who accepts the contract and subsequently installs a positive service capacity will not be worse off when he is weakly risk-seeking ($\overline{\eta} \in (0, 8/9)$) than risk-neutral ($\overline{\eta} = 0$).*

We return to the case of $\overline{\eta} \geq 8/9$ in Sects. 5.2.1 and 5.3.

5.1.2 Principal's Optimal Strategy

We now proceed to derive the principal's optimal strategy. Anticipating the agent's optimal selection of $\mu^*(w,p)$ the principal chooses w and p that maximize her expected profit rate by solving the optimization problem

$$\max_{w>0,p>0} \Pi_P(w,p;\mu^*(w,p)) = \max_{w>0,p>0} \left\{ \frac{r\mu^*(w,p)}{\lambda+\mu^*(w,p)} - w + \frac{p\lambda}{\lambda+\mu^*(w,p)} \right\} \tag{5.11}$$

Denote $(w^*,p^*) = \text{argmax}_{w>0,p>0}\, \Pi_P(w,p;\mu^*(w,p))$.

Before deriving the principal's optimal strategy, we examine the case when the principal offers $p = \overline{p}_2$ and $w \geq \overline{w}_2$, under which the agent is indifferent about installing two different service capacities. In such a case, the solution $((w,p),\mu)$ has to be an admissible solution (see Definition 5.1). We state this case formally in Proposition 5.13.

Proposition 5.13. *Suppose a weakly risk-seeking agent. Assume that the principal's possible offers are constrained to set $\{(w,p) : p = \overline{p}_2, w \geq \overline{w}_2\}$.*

(a) If $r \in (0,\overline{p}_2)$, then the agent installs $\mu^* = \sqrt{(1-\overline{\eta})\overline{p}_2\lambda} - \lambda$ if offered a contract.
(b) If $r = \overline{p}_2$, then both $\mu^* = \sqrt{(1-\overline{\eta})\overline{p}_2\lambda} - \lambda$ and $\mu^* = \sqrt{\overline{p}_2\lambda} - \lambda$ lead to admissible solutions and the agent installs either $\sqrt{(1-\overline{\eta})\overline{p}_2\lambda} - \lambda$ or $\sqrt{\overline{p}_2\lambda} - \lambda$ if offered a contract.
(c) If $r > \overline{p}_2$, then the agent installs $\mu^* = \sqrt{\overline{p}_2\lambda} - \lambda$ if offered a contract.

Proof. Note that for $w \geq \overline{w}_2$ we have $\partial\Pi_P(w,\overline{p}_2;\mu)/\partial\mu = (r-\overline{p}_2)\lambda/(\lambda+\mu)^2$. Define $\mu_L \equiv \sqrt{(1-\overline{\eta})\overline{p}_2\lambda} - \lambda$ and $\mu_H \equiv \sqrt{\overline{p}_2\lambda} - \lambda$. Note that $\mu_H > \mu_L$. If $r \in (0,\overline{p}_2)$, then $\partial\Pi_P/\partial\mu < 0$, therefore $((w,\overline{p}_2),\mu_L) \succeq ((w,\overline{p}_2),\mu_H)$. If the principal offers a contract (the conditions are discussed in detail in Theorem 5.17 that follows), then by Definition 5.1 only μ_L leads to admissible solutions and we obtain (a). If $r > \overline{p}_2$, then $\partial\Pi_P/\partial\mu > 0$, therefore $((w,\overline{p}_2),\mu_H) \succeq ((w,\overline{p}_2),\mu_L)$.

If the principal offers a contract (the conditions are discussed in Theorem 5.17 that follows), then by Definition 5.1 only μ_H leads to admissible solutions and we obtain (c). If $r = \overline{p}_2$, then $\partial \Pi_P / \partial \mu = 0$, indicating that the principal receives the same expected profit rate when the agent installs capacity μ_L or μ_H. If the principal offers a contract (the conditions are discussed in Theorem 5.17 that follows), then both μ_L and μ_H lead to admissible solutions. Therefore we obtain (b). □

Notation:

$$\overline{r}_1 \equiv \overline{\eta}\overline{p}_2 + (1-\overline{\eta})\sqrt{\overline{p}_1\overline{p}_2} - \frac{\overline{\eta}\overline{p}_2}{2}\left(\frac{\sqrt{\overline{p}_2}}{\sqrt{\overline{p}_2}-\sqrt{\overline{p}_1}}\right), \overline{r}_2 \equiv (1-\overline{\eta})\overline{p}_2 + \overline{\eta}\overline{p}_2\left(\frac{\sqrt{\overline{p}_2}-\sqrt{\overline{p}_1}}{\sqrt{\overline{p}_1}}\right) \tag{5.12}$$

Note that $\overline{r}_1$ and $\overline{r}_2$ are functions of λ and $\overline{\eta}$. However we suppress the parameters $(\lambda, \overline{\eta})$.

We define $\overline{p}_{cu}$ as follows[1]:

$$\overline{p}_{cu} \equiv \frac{1}{9a^2}\left(b + C + \overline{C}\right)^2 \tag{5.13}$$

where $a \equiv \overline{\eta}$, $b \equiv (1-2\overline{\eta})\sqrt{\overline{p}_1}$, and $d \equiv -r\sqrt{\overline{p}_1}$ and

$$C \equiv \sqrt[3]{\frac{\Delta_1 + \sqrt{\Delta_1^2 - 4\Delta_0^3}}{2}}, \overline{C} \equiv \sqrt[3]{\frac{\Delta_1 - \sqrt{\Delta_1^2 - 4\Delta_0^3}}{2}}, \text{ where } \Delta_0 \equiv b^2,\ \Delta_1 \equiv 2b^3 + 27a^2 d$$

Replacing Δ_0 and Δ_1 by the expressions of a, b and d we have

$$C = \sqrt[3]{\frac{2(1-2\overline{\eta})^3\sqrt{\overline{p}_1^3} - 27\overline{\eta}^2 r\sqrt{\overline{p}_1} + \sqrt{-108\overline{\eta}^2 r(1-2\overline{\eta})^3\overline{p}_1^2 + 729\overline{\eta}^4 r^2\overline{p}_1}}{2}} \text{ and}$$

$$\overline{C} = \sqrt[3]{\frac{2(1-2\overline{\eta})^3\sqrt{\overline{p}_1^3} - 27\overline{\eta}^2 r\sqrt{\overline{p}_1} - \sqrt{-108\overline{\eta}^2 r(1-2\overline{\eta})^3\overline{p}_1^2 + 729\overline{\eta}^4 r^2\overline{p}_1}}{2}}$$

Next we state a number of technical lemmas (see proofs in the Appendix).

Lemma 5.14. *Let* $8/9 > \overline{\eta} > 0$ *and* $\lambda > 0$, *then*

[1]The subscript "*cu*" stands for "cubic" because (5.13) is the square of the solution to Eq. (A.2), which is a cubic equation that is introduced later in the proof for Lemma 5.15.

(a) $\overline{p}_2 > (1-\overline{\eta})\overline{p}_2 + \overline{\eta}\overline{p}_2\left(\frac{\sqrt{\overline{p}_2}-\sqrt{\overline{p}_1}}{\sqrt{\overline{p}_1}}\right) > \overline{\eta}\overline{p}_2 + (1-\overline{\eta})\sqrt{\overline{p}_1\overline{p}_2} - \frac{\overline{\eta}\overline{p}_2}{2}\left(\frac{\sqrt{\overline{p}_2}}{\sqrt{\overline{p}_2}-\sqrt{\overline{p}_1}}\right).$

(b) $(1-\overline{\eta})\overline{p}_2 + \overline{\eta}\overline{p}_2\left(\frac{\sqrt{\overline{p}_2}-\sqrt{\overline{p}_1}}{\sqrt{\overline{p}_1}}\right) > \lambda.$

Lemma 5.15. *Consider* $\max_{x\in[\sqrt{\overline{p}_1},\sqrt{\overline{p}_2}]} f(x)$ *where* $f(x) = r + \lambda - \overline{\eta}x^2/2 - \sqrt{\overline{p}_1}\left((1-2\overline{\eta})x + r/x\right)$ *and denote* $x^* = \operatorname{argmax}_{x\in[\sqrt{\overline{p}_1},\sqrt{\overline{p}_2}]} f(x)$. *The solutions to this optimization problem are:*

(a) $x^* = \sqrt{\overline{p}_1}$ *if* $r \in (0,\lambda]$.
(b) $x^* = \sqrt{\overline{p}_{cu}} \in \left(\sqrt{\overline{p}_1}, \sqrt{\overline{p}_2}\right)$ *if* $r \in (\lambda, \overline{r}_2)$.
(c) $x^* = \sqrt{\overline{p}_2}$ *if* $r \geq \overline{r}_2$.

Lemma 5.16. *Consider* $\max_{x\geq\sqrt{\overline{p}_2}} f(x)$ *where* $f(x) = r + \lambda - \sqrt{\lambda}\,(x + r/x)$ *and denote* $x^* = \operatorname{argmax}_{x\geq\sqrt{\overline{p}_2}} f(x)$. *Solutions to this optimization problem are*

(a) $x^* = \sqrt{\overline{p}_2}$ *if* $r \in (0,\overline{p}_2]$.
(b) $x^* = \sqrt{r}$ *if* $r > \overline{p}_2$.

Lemma 5.14 implies $\overline{p}_2 > \overline{r}_2 > \overline{r}_1$ and $\overline{r}_2 > \lambda$.

Recall that Proposition 5.10 describes the agent's optimal response to each pair $(w,p) \in \mathbb{R}^2_+$. Since the principal will not propose a contract that ex ante is going to be rejected by a weakly risk-seeking (WRS) agent, therefore Theorem 5.17 only considers pairs $(w,p) \in \mathbb{R}^2_+$ that result in agent's non-negative expected utility rate. Define

$$\begin{aligned}
\mathfrak{D}_{(5.7)} &\equiv \{(w,p) \text{ that satisfies (5.7) when } \overline{\eta} \in (0,8/9)\}\\
\mathfrak{D}_{(5.8)} &\equiv \{(w,p) \text{ that satisfies (5.8) when } \overline{\eta} \in (0,8/9)\}\\
\mathfrak{D}_{(5.9)} &\equiv \{(w,p) \text{ that satisfies (5.9) when } \overline{\eta} \in (0,8/9)\}\\
\mathfrak{D}_{(5.10)} &\equiv \{(w,p) \text{ that satisfies (5.10) when } \overline{\eta} \in (0,8/9)\}\\
\mathfrak{D}_{\text{WRS}} &\equiv \mathfrak{D}_{(5.7)} \cup \mathfrak{D}_{(5.8)} \cup \mathfrak{D}_{(5.9)} \cup \mathfrak{D}_{(5.10)}
\end{aligned} \tag{5.14}$$

Theorem 5.17. *Given a weakly risk-seeking agent and* $(w,p) \in \mathfrak{D}_{\text{WRS}}$.

(a) *If* $r \in (0,\lambda]$, *then the principal does not propose a contract.*
(b) *If* $r \in (\lambda, \overline{r}_2)$, *then the principal's offer and the capacity installed by the agent are:*

$$(w^*,p^*) = \left(\frac{\overline{\eta}\overline{p}_{cu}}{2} + 2\sqrt{(1-\overline{\eta})\overline{p}_{cu}\lambda} - \lambda, \overline{p}_{cu}\right) \text{ and } \mu^*(w^*,p^*) = \sqrt{(1-\overline{\eta})\overline{p}_{cu}\lambda} - \lambda \tag{5.15}$$

and the principal's expected profit rate is:

$$\Pi_P(w^*, p^*; \mu^*(w^*, p^*)) = r + \lambda - \frac{\overline{\eta}\overline{p}_{cu}}{2} - \sqrt{\overline{p}_1}\left((1 - 2\overline{\eta})\sqrt{\overline{p}_{cu}} + \frac{r}{\sqrt{\overline{p}_{cu}}}\right) \tag{5.16}$$

(c) *If $r \in [\overline{r}_2, \overline{p}_2]$, then the principal's offer and the capacity installed by the agent are:*

$$\left(w^*, p^*\right) = (\overline{w}_2, \overline{p}_2) \text{ and } \mu^*(w^*, p^*) = \sqrt{(1 - \overline{\eta})\overline{p}_2\lambda} - \lambda \tag{5.17}$$

and the principal's expected profit rate is:

$$\Pi_P(w^*, p^*; \mu^*(w^*, p^*)) = r + \lambda - \frac{\overline{\eta}\overline{p}_2}{2} - \sqrt{\overline{p}_1}\left((1 - 2\overline{\eta})\sqrt{\overline{p}_2} + \frac{r}{\sqrt{\overline{p}_2}}\right) \tag{5.18}$$

(d) *If $r > \overline{p}_2$, then the principal's offer and the capacity installed by the agent are*

$$\left(w^*, p^*\right) = \left(2\sqrt{r\lambda} - \lambda, r\right) \text{ and } \mu^*(w^*, p^*) = \sqrt{r\lambda} - \lambda \tag{5.19}$$

and the principal's expected profit rate is:

$$\Pi_P(w^*, p^*; \mu^*(w^*, p^*)) = r - 2\sqrt{r\lambda} + \lambda \tag{5.20}$$

Proof. The structure of the proof for Theorem 5.17 is depicted in Fig. 5.7.

Case $(w, p) \in \mathfrak{D}_{(5.7)}$: According to Proposition 5.10 part (a), in case the principal makes an offer, the agent accepts the contract but does not install any service capacity. Since $\partial\Pi_P/\partial w = -1 < 0$, thus we have $w^* = (1 - \overline{\eta}/2)p$ and from (3.3) $\Pi_P(w^*, p; \mu^*(w^*, p)) = -w^* + p = \overline{\eta}p/2 > 0$. However in such case $p > w^* = (1 - \overline{\eta}/2)p$, which violates condition (c) in Definition 5.1, therefore $((w^* = (1 - \overline{\eta}/2)p, p), \mu^* = 0)$ is not an admissible solution and the principal does not propose a contract.

Case $(w, p) \in \mathfrak{D}_{(5.8)} \cup \mathfrak{D}_{(5.9)}$: According to Proposition 5.10 part (b), if $(w, p) \in \mathfrak{D}_{(5.8)}$, then in case the principal makes an offer, the agent accepts the contract and installs $\sqrt{(1 - \overline{\eta})p\lambda} - \lambda$. Since $\partial\Pi_P/\partial w = -1 < 0$, therefore $w^* = \overline{\eta}p/2 + 2\sqrt{(1 - \overline{\eta})p\lambda} - \lambda$. According to Propositions 5.10 part (c) and 5.13, if $(w, p) \in \mathfrak{D}_{(5.9)}$ (which implies $p = \overline{p}_2$), then in case the principal makes an offer, the agent accepts the contract and installs $\sqrt{(1 - \overline{\eta})\overline{p}_2\lambda} - \lambda$ if $r \in (0, \overline{p}_2)$, installs either $\sqrt{(1 - \overline{\eta})\overline{p}_2\lambda} - \lambda$ or $\sqrt{\overline{p}_2\lambda} - \lambda$ if $r = \overline{p}_2$, or installs $\sqrt{\overline{p}_2\lambda} - \lambda$ if $r > \overline{p}_2$. Since $\partial\Pi_P/\partial w = -1 < 0$, therefore $w^* = \overline{w}_2$. For convenience denote the principal's expected profit rate when $(w, p) = (\overline{w}_2, \overline{p}_2)$ and $\mu^*(w, p) = \sqrt{(1 - \overline{\eta})\overline{p}_2\lambda} - \lambda$ by

Fig. 5.7 Structure of the proof for Theorem 5.17

$\Pi_P^L(\overline{p}_2)$, and denote the principal's expected profit rate when $(w, p) = (\overline{w}_2, \overline{p}_2)$ and $\mu^*(w, p) = \sqrt{\overline{p}_2\lambda} - \lambda$ by $\Pi_P^H(\overline{p}_2)$. By plugging the value of w, p and μ into (3.3):

$$\Pi_P^L(\overline{p}_2) = r + \lambda - \frac{\overline{\eta}\overline{p}_2}{2} - \sqrt{\overline{p}_1}\left((1 - 2\overline{\eta})\sqrt{\overline{p}_2} + \frac{r}{\sqrt{\overline{p}_2}}\right) = \left(\frac{\sqrt{\overline{p}_2} - \sqrt{\overline{p}_1}}{\sqrt{\overline{p}_2}}\right)(r - \overline{r}_1) \tag{5.21}$$

$$\Pi_P^H(\overline{p}_2) = r + \lambda - \sqrt{\lambda}\left(\sqrt{\overline{p}_2} + \frac{r}{\sqrt{\overline{p}_2}}\right) \tag{5.22}$$

In such case the principal's optimization problem is $\max_{p \in [\overline{p}_1, \overline{p}_2]} \Pi_P(w^*, p; \mu^*(w^*, p))$ where:

$$\Pi_P(w^*, p; \mu^*(w^*, p)) = \begin{cases} r + \lambda - \dfrac{\overline{\eta}p}{2} - \sqrt{\overline{p}_1}\left((1 - 2\overline{\eta})\sqrt{p} + \dfrac{r}{\sqrt{p}}\right), \text{ for } p \in [\overline{p}_1, \overline{p}_2) \\ \max\left\{\Pi_P^L(\overline{p}_2), \Pi_P^H(\overline{p}_2)\right\}, \text{ for } p = \overline{p}_2 \end{cases}$$

Define $x \equiv \sqrt{p}$, the expression $r + \lambda - \overline{\eta}p/2 - \sqrt{\overline{p}_1}\,((1 - 2\overline{\eta})\sqrt{p} + r/\sqrt{p})$ can be restated as $f(x) = r + \lambda - \overline{\eta}x^2/2 - \sqrt{\overline{p}_1}\,((1 - 2\overline{\eta})x + r/x)$. Maximizing $f(x)$

with respect to x over $\left[\sqrt{\overline{p}_1}, \sqrt{\overline{p}_2}\right]$ is equivalent to maximizing $r + \lambda - \overline{\eta}p/2 - \sqrt{\overline{p}_1}\left((1-2\overline{\eta})\sqrt{p} + r/\sqrt{p}\right)$ with respect to p over $[\overline{p}_1, \overline{p}_2]$ in the sense that

$$\underset{p\in[\overline{p}_1,\overline{p}_2]}{\operatorname{argmax}}\left\{r + \lambda - \frac{\overline{\eta}p}{2} - \sqrt{\overline{p}_1}\left((1-2\overline{\eta})\sqrt{p} + \frac{r}{\sqrt{p}}\right)\right\} = \left(\underset{x\in\left[\sqrt{\overline{p}_1},\sqrt{\overline{p}_2}\right]}{\operatorname{argmax}} f(x)\right)^2$$

From Lemma 5.14, $\overline{p}_2 > \overline{r}_2 > \lambda$ and we examine the following subcases:

Subcase $r \in (0, \lambda]$: According to Lemma 5.15 part (a), $p^* = \overline{p}_1$, which is covered in $(w, p) \in \mathfrak{D}_{(5.7)}$ and the principal does not propose a contract.

Subcase $r \in (\lambda, \overline{r}_2)$: According to Lemma 5.15 part (b), $p^* = \overline{p}_{cu}$ and the principal's expected profit rate is $\Pi_P(w^*, p^*; \mu^*(w^*, p^*)) > \Pi_P((1 - \overline{\eta}/2)\overline{p}_1, \overline{p}_1; 0) = \overline{\eta}\overline{p}_1/2 > 0$. Therefore the principal proposes a contract with $w^* = \overline{\eta}\overline{p}_{cu}/2 + 2\sqrt{(1-\overline{\eta})\overline{p}_{cu}\lambda} - \lambda$ and $p^* = \overline{p}_{cu}$ that induces the agent to install $\mu^*(w^*, p^*) = \sqrt{(1-\overline{\eta})\overline{p}_{cu}\lambda} - \lambda$.

Subcase $r \in [\overline{r}_2, \overline{p}_2]$: According to Lemma 5.15 part (c), $p^* = \overline{p}_2$ and according to Proposition 5.13 part (a) and (b) the principal's expected profit rate is $\Pi_P(w^*, p^*; \mu^*(w^*, p^*)) = \Pi_P^L(\overline{p}_2) > \Pi_P((1 - \overline{\eta}/2)\overline{p}_1, \overline{p}_1; 0) = \overline{\eta}\overline{p}_1/2 > 0$. Therefore the principal proposes a contract with $w^* = \overline{w}_2$ and $p^* = \overline{p}_2$ that induces the agent to install $\mu^*(w^*, p^*) = \sqrt{(1-\overline{\eta})\overline{p}_2\lambda} - \lambda$.

Subcase $r > \overline{p}_2$: According to Lemma 5.15 part (c), $p^* = \overline{p}_2$ and according to Proposition 5.13 part (c) her expected profit rate is $\Pi_P(w^*, p^*; \mu^*(w^*, p^*)) = \Pi_P^H(\overline{p}_2) > \Pi_P^L(\overline{p}_2) > \Pi_P((1 - \overline{\eta}/2)\overline{p}_1, \overline{p}_1; 0) = \overline{\eta}\overline{p}_1/2 > 0$. Therefore the principal proposes a contract with $w^* = \overline{w}_2$ and $p^* = \overline{p}_2$ that induces the agent to install $\mu^*(w^*, p^*) = \sqrt{\overline{p}_2\lambda} - \lambda$.

Case $(w, p) \in \mathfrak{D}_{(5.9)} \cup \mathfrak{D}_{(5.10)}$: According to Proposition 5.10 part (d), if $(w, p) \in \mathfrak{D}_{(5.10)}$, then in case the principal makes an offer, the agent accepts the contract and installs $\sqrt{p\lambda} - \lambda$. Since $\partial\Pi_P/\partial w = -1 < 0$, therefore $w^* = 2\sqrt{p\lambda} - \lambda$. According to Propositions 5.10 part (c) and 5.13, if $(w, p) \in \mathfrak{D}_{(5.9)}$ (which implies $p = \overline{p}_2$), then in case the principal makes an offer, the agent accepts the contract and installs $\sqrt{(1-\overline{\eta})\overline{p}_2\lambda} - \lambda$ if $r \in (0, \overline{p}_2)$, installs either $\sqrt{(1-\overline{\eta})\overline{p}_2\lambda} - \lambda$ or $\sqrt{\overline{p}_2\lambda} - \lambda$ if $r = \overline{p}_2$, or installs $\sqrt{\overline{p}_2\lambda} - \lambda$ if $r > \overline{p}_2$. Since $\partial\Pi_P/\partial w = -1 < 0$, therefore $w^* = \overline{w}_2$. Recall the definition of $\Pi_P^L(\overline{p}_2)$ and $\Pi_P^H(\overline{p}_2)$ (see Eqs. (5.21) and (5.22)). Thus the principal's optimization problem is $\max_{p\geq\overline{p}_2} \Pi_P(w^*, p; \mu^*(w^*, p))$ where:

$$\Pi_P(w^*, p; \mu^*(w^*, p)) = \begin{cases} \max\left\{\Pi_P^L(\overline{p}_2), \Pi_P^H(\overline{p}_2)\right\}, \text{ for } p = \overline{p}_2 \\ r + \lambda - \sqrt{\lambda}\left(\sqrt{p} + \dfrac{r}{\sqrt{p}}\right), \text{ for } p > \overline{p}_2 \end{cases}$$

Define $x \equiv \sqrt{p}$, the expression $r + \lambda - \sqrt{\lambda}\left(\sqrt{p} + r/\sqrt{p}\right)$ can be restated as $f(x) = r + \lambda - \sqrt{\lambda}\,(x + r/x)$. Maximizing $f(x)$ with respect to $x \geq \sqrt{\overline{p}_2}$ is equivalent to maximizing $r + \lambda - \sqrt{\lambda}\left(\sqrt{p} + r/\sqrt{p}\right)$ with respect to $p \geq \overline{p}_2$ in the sense that

$$\underset{p \geq \overline{p}_2}{\operatorname{argmax}} \left\{ r + \lambda - \sqrt{\lambda}\left(\sqrt{p} + \frac{r}{\sqrt{p}}\right) \right\} = \left(\underset{x \geq \sqrt{\overline{p}_2}}{\operatorname{argmax}} f(x) \right)^2$$

According to Lemma 5.14, $\overline{p}_2 > \overline{r}_1$. Also note that $\lim_{\overline{\eta} \to 0+} \overline{r}_1 = 2\lambda$ and according to Lemma 5.8 $\lim_{\overline{\eta} \to 8/9-} \overline{r}_1 = -\infty$. Therefore we examine the following subcases:

Subcase $r \in (0, \max\{0, \overline{r}_1\}]$: According to Lemma 5.16 part (a), $p^* = \overline{p}_2$. According to Proposition 5.13 part (a), $\Pi_P(w^*, p^*; \mu^*(w^*, p^*)) = \Pi_P^L(\overline{p}_2) \leq 0$, therefore the principal does not propose a contract.

Subcase $r \in (\max\{0, \overline{r}_1\}, \overline{p}_2]$: According to Lemma 5.16 part (a), $p^* = \overline{p}_2$. According to Proposition 5.13 part (a) and (b), $\Pi_P(w^*, p^*; \mu^*(w^*, p^*)) = \Pi_P^L(\overline{p}_2) > 0$, therefore the principal proposes a contract with $w^* = \overline{w}_2$ and $p^* = \overline{p}_2$ that induces the agent to install $\mu^*(w^*, p^*) = \sqrt{(1-\overline{\eta})\overline{p}_2\lambda} - \lambda$.

Subcase $r > \overline{p}_2$: According to Proposition 5.16 part (b), $p^* = r$ and the principal's expected profit rate is $\Pi_P(w^*, p^*; \mu^*(w^*, p^*)) = r - 2\sqrt{r\lambda} + \lambda > 0$. Thus the principal proposes a contract with $w^* = 2\sqrt{r\lambda} - \lambda$ and $p^* = r$ that induces the agent to install $\mu^*(w^*, p^*) = \sqrt{r\lambda} - \lambda$.

To summarize, if $r \in (0, \lambda]$, then the principal does not propose a contract. If $r \in (\lambda, \overline{r}_2)$, then the principal offers $(w^*, p^*) = \left(\overline{\eta}\overline{p}_{cu}/2 + 2\sqrt{(1-\overline{\eta})\overline{p}_{cu}\lambda} - \lambda, \overline{p}_{cu}\right)$ and the agent installs capacity $\mu^*(w^*, p^*) = \sqrt{(1-\overline{\eta})\overline{p}_{cu}\lambda} - \lambda$. If $r \in [\overline{r}_2, \overline{p}_2]$, then the principal offers $(w^*, p^*) = (\overline{w}_2, \overline{p}_2)$ and the agent installs capacity $\mu^*(w^*, p^*) = \sqrt{(1-\overline{\eta})\overline{p}_2\lambda} - \lambda$. If $r > \overline{p}_2$, then the principal offers $(w^*, p^*) = \left(2\sqrt{r\lambda} - \lambda, r\right)$ and the agent installs capacity $\mu^*(w^*, p^*) = \sqrt{r\lambda} - \lambda$. □

Theorem 5.17 indicates that the existence of a beneficial contract for weakly risk-seeking agent is determined exogenously by r, λ, and $\overline{\eta}$.

5.2 Optimal Strategies for the Moderately Risk-Seeking Agent

For the moderately risk-seeking agent we first derive the agent's optimal strategy. The agent's optimization problem is defined in (5.3).

Notation:

$$\overline{p}_3 \equiv \frac{2\lambda}{2 + \overline{\eta} - 2\sqrt{2\overline{\eta}}} \tag{5.23}$$

and the following identity is verified using the definition of $\overline{p}_3$:

$$\overline{w}_3 \equiv \left(1 - \frac{\overline{\eta}}{2}\right)\overline{p}_3 = 2\sqrt{\overline{p}_3\lambda} - \lambda \tag{5.24}$$

Note that $\overline{p}_3$ and $\overline{w}_3$ are functions of λ and $\overline{\eta}$. However we suppress the parameters $(\lambda, \overline{\eta})$.

Lemma 5.18. *Let $2 > \overline{\eta} \geq 8/9$ and $\lambda > 0$, then $2\lambda / \left(2 + \overline{\eta} - 2\sqrt{2\overline{\eta}}\right) > 4\lambda$ (see proof in the Appendix).*

We describe a moderately risk-seeking agent's optimal response to any $(w, p) \in \mathbb{R}^2_+$ in Proposition 5.19.

Proposition 5.19. *Consider a moderately risk-seeking agent ($\overline{\eta} \in [8/9, 2)$).*

(a) *Given*

$$p \in (0, \overline{p}_3) \text{ and } w \geq \left(1 - \frac{\overline{\eta}}{2}\right)p \tag{5.25}$$

then the agent accepts the contract and installs $\mu^*(w,p) = 0$ with resulting expected utility rate $u_A(\mu^*(w,p); w, p) = w - (1 - \overline{\eta}/2)p \geq 0$. The agent rejects the contract if $p \in (0, \overline{p}_3]$ and $w \in (0, (1 - \overline{\eta}/2)p)$.

(b) *Given*

$$p = \overline{p}_3 \text{ and } w \geq \overline{w}_3 \tag{5.26}$$

then the agent accepts the contract and is indifferent installing either $\mu^*(w,p) = 0$ or $\mu^*(w,p) = \sqrt{\overline{p}_3\lambda} - \lambda$. In both cases the agent's expected utility rate is $u_A(\mu^*(w,p); w, p) = w - (1 - \overline{\eta}/2)\overline{p}_3 = w - 2\sqrt{\overline{p}_3\lambda} + \lambda \geq 0$. If $r \in (0, \overline{p}_3]$, then neither $\mu^* = 0$ nor $\mu^* = \sqrt{\overline{p}_3\lambda} - \lambda$ leads to admissible solutions (see Definition 5.1). If $r > \overline{p}_3$, then there exists w^* such that $\left((w^*, \overline{p}_3), \mu^* = \sqrt{\overline{p}_3\lambda} - \lambda\right)$ is the only admissible solution (for proof see Proposition 5.20). He rejects the contract if $p = \overline{p}_3$ and $w \in (0, \overline{w}_3)$.

(c) *Given*

$$p > \overline{p}_3 \text{ and } w \geq 2\sqrt{p\lambda} - \lambda \tag{5.27}$$

then the agent accepts the contract and installs $\mu^*(w,p) = \sqrt{p\lambda} - \lambda$ with resulting expected utility rate $u_A(\mu^*(w,p); w, p) = w - 2\sqrt{p\lambda} + \lambda \geq 0$. The agent rejects the contract if $p > \overline{p}_3$ and $w \in \left(0, 2\sqrt{p\lambda} - \lambda\right)$.

Proof. According to Table 5.1, the optimization of $u(\mu)$ when $\overline{\eta} \in [8/9, 1)$ versus $\overline{\eta} \in [1, 2)$ is different. Therefore we prove the proposition separately for $\overline{\eta} \in [8/9, 1)$ and $\overline{\eta} \in [1, 2)$.

Case $\overline{\eta} \in [8/9, 1)$: According to Lemmas 5.6 and 5.18, $4\overline{p}_1 > \overline{p}_1 \geq \overline{p}_3 > 4\lambda$. Figure 5.8 shows the shape of $u(\mu)$ when $\overline{\eta} \in [8/9, 1)$ and the value of p falls in different ranges. The structure of the proof when $\overline{\eta} \in [8/9, 1)$ is depicted in Fig. 5.9.

Case $p \in (0, 4\lambda]$: According to Table 5.1, $u(\mu)$ is decreasing with respect to $\mu \geq 0$. Therefore the agent's optimal service capacity is $\mu^*(w, p) = 0$ and from Eq. (5.3) $u(\mu^*(w, p)) = w - (1 - \overline{\eta}/2)p$. Note that $1 - \overline{\eta}/2 > 0$.

Case $w \in \left(0, (1 - \overline{\eta}/2)p\right)$: $u(\mu^*(w, p)) < 0$, therefore the agent rejects the contract.

Case $w \geq (1 - \overline{\eta}/2)p$: $u(\mu^*(w, p)) \geq 0$, thus the agent would accept the contract if offered.

Case $p \in \left(4\lambda, \overline{p}_1\right]$: According to Table 5.1, there is a service capacity that maximizes $u(\mu)$ for $\mu \in [0, \lambda)$ and a service capacity that maximizes $u(\mu)$ for $\mu > \lambda$. Denote the optimal service capacity in $[0, \lambda)$ by $\mu^*_{[0,\lambda)}(w, p)$. Since $u(\mu)$ is decreasing with respect to μ over $[0, \lambda)$, therefore $\mu^*_{[0,\lambda)}(w, p) = 0$ and from (5.3) $u\left(\mu^*_{[0,\lambda)}(w, p)\right) = w - (1 - \overline{\eta}/2)p$. Denote the optimal service capacity for $\mu > \lambda$ by $\mu^*_{\lambda}(w, p)$. From first order condition $\mu^*_{\lambda}(w, p) = \sqrt{p\lambda} - \lambda$ and from (5.3) $u\left(\mu^*_{\lambda}(w, p)\right) = w - 2\sqrt{p\lambda} + \lambda$. The agent has to choose one of the two service capacities and he installs the one with higher expected utility rate. Note that $u\left(\mu^*_{\lambda}(w, p)\right) - u\left(\mu^*_{[0,\lambda)}(w, p)\right) = (1 - \overline{\eta}/2)p - 2\sqrt{p\lambda} + \lambda$. According to Lemma 5.5, $4\lambda > 2\lambda/\left(2 + \overline{\eta} + 2\sqrt{2\overline{\eta}}\right)$. According to Lemmas 5.6 and 5.18, $\overline{p}_1 \geq \overline{p}_3 > 4\lambda$, therefore we examine the following subcases.

Subcase $p \in \left(4\lambda, \overline{p}_3\right)$: By Lemma 5.4 part (a), $u\left(\mu^*_{[0,\lambda)}(w, p)\right) > u\left(\mu^*_{\lambda}(w, p)\right)$, therefore the agent's optimal service capacity is $\mu^*(w, p) = 0$ and $u(\mu^*(w, p)) = w - (1 - \overline{\eta}/2)p$. Note that $1 - \overline{\eta}/2 > 0$.

Subsubcase $w \in \left(0, (1 - \overline{\eta}/2)p\right)$: $u(\mu^*(w, p)) < 0$, therefore the agent rejects the contract.

Subsubcase $w \geq (1 - \overline{\eta}/2)p$: $u(\mu^*(w, p)) \geq 0$, thus the agent would accept the contract if offered.

Subcase $p = \overline{p}_3$: According to Lemma 5.4 part (c), $u\left(\mu^*_{[0,\lambda)}(w, p)\right) = u\left(\mu^*_{\lambda}(w, p)\right)$, indicating that installing $\mu^*_{[0,\lambda)}(w, \overline{p}_3)$ or $\mu^*_{\lambda}(w, \overline{p}_3)$ leads to the same agent's expected utility rate. Therefore the agent is indifferent about installing $\mu^*(w, p) = 0$ or $\mu^*(w, p) = \sqrt{p\lambda} - \lambda$. Still, the capacity value has to lead to admissible solutions (see Proposition 5.20). Recall that by definition $\overline{w}_3 = (1 - \overline{\eta}/2)\overline{p}_3$ (see (5.24)). Note that $1 - \overline{\eta}/2 > 0$.

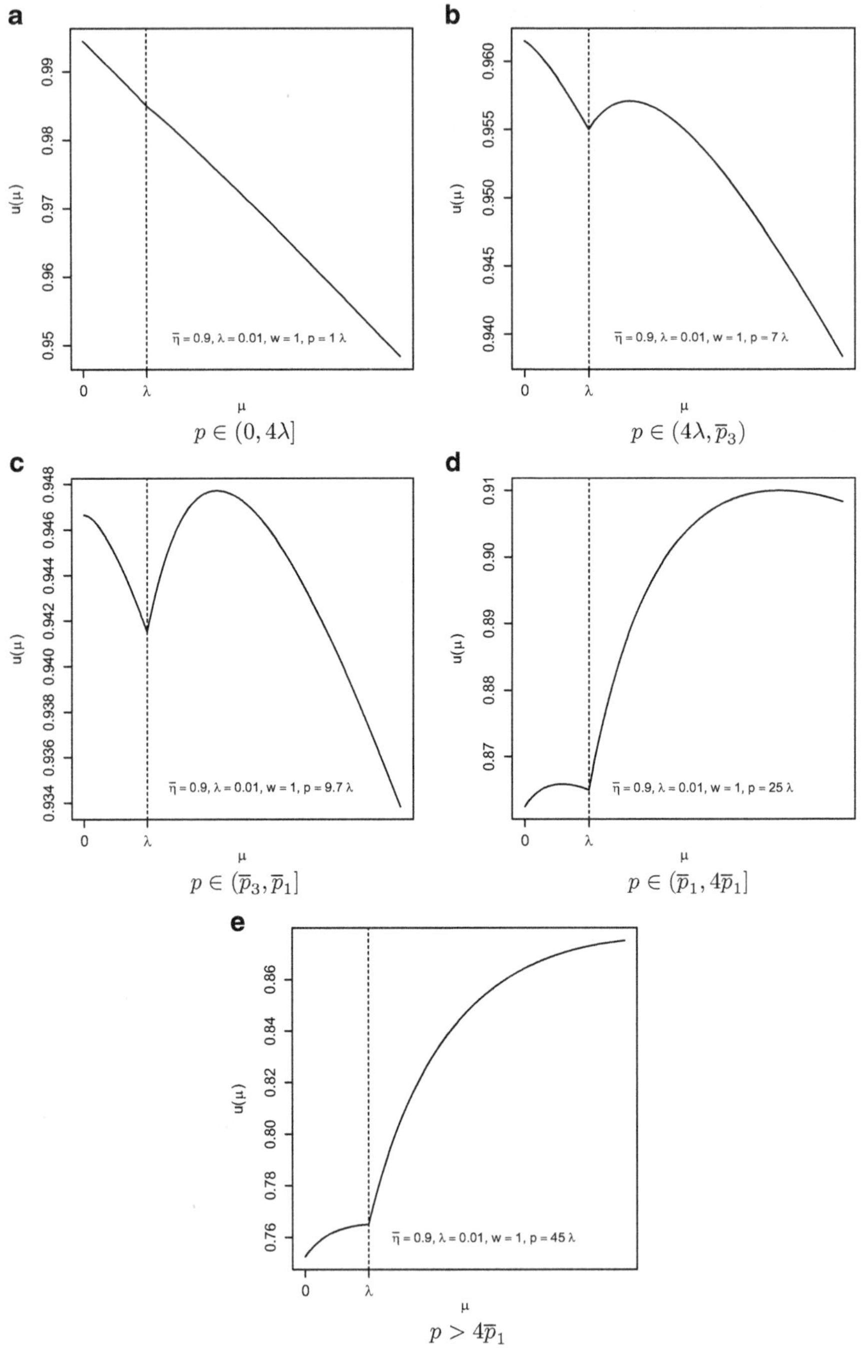

Fig. 5.8 Illustration of the forms of $u(\mu)$ when $\overline{\eta} \in [8/9, 1)$

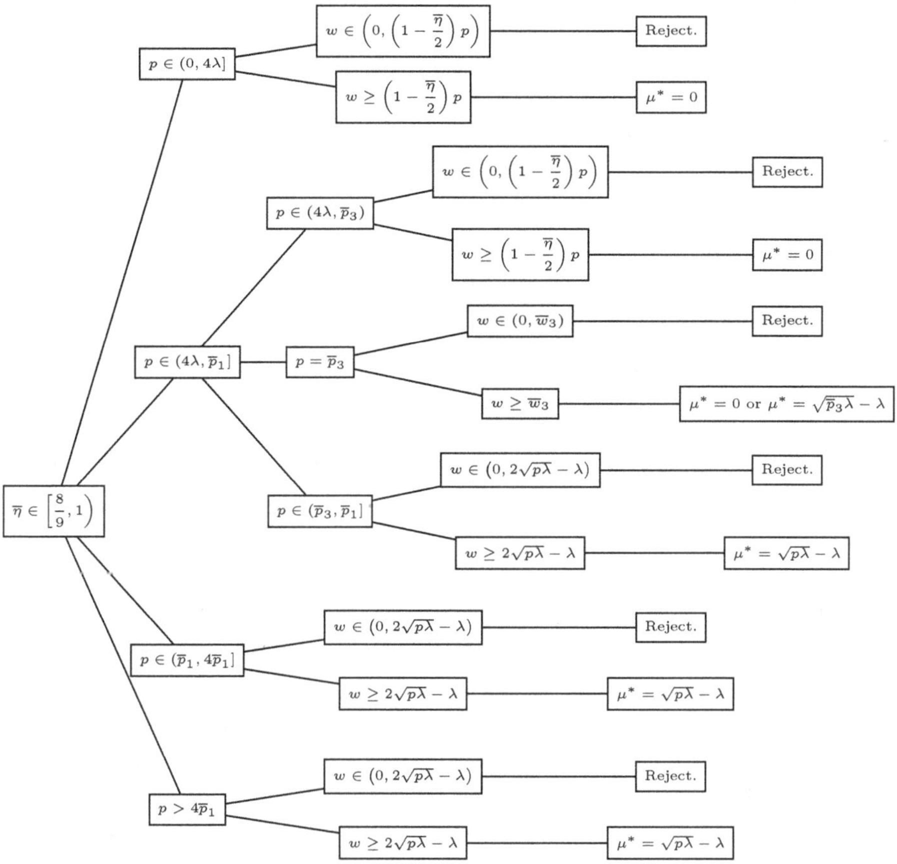

Fig. 5.9 Structure of the proof for Proposition 5.19 when $\overline{\eta} \in [8/9, 1)$

Subsubcase $w \in \left(0, (1 - \overline{\eta}/2)\overline{p}_3\right)$: $u(\mu^*(w, p)) < 0$, therefore the agent rejects the contract.

Subsubcase $w \geq (1 - \overline{\eta}/2)\overline{p}_3$: $u(\mu^*(w, p)) \geq 0$, therefore the agent would accept the contract if offered.

Subcase $p \in \left(\overline{p}_3, \overline{p}_1\right]$: According to Lemma 5.4 part (b), $u\left(\mu^*_{\lambda}(w, p)\right) > u\left(\mu^*_{[0,\lambda)}(w, p)\right)$, therefore the agent's optimal service capacity is $\mu^*(w, p) = \sqrt{p\lambda} - \lambda$ and $u(\mu^*(w, p)) = w - 2\sqrt{p\lambda} + \lambda$. Since $p > \overline{p}_3 > 4\lambda \Rightarrow 2\sqrt{p\lambda} - \lambda > 3\lambda > 0$, therefore we examine the following subcases.

Subsubcase $w \in \left(0, 2\sqrt{p\lambda} - \lambda\right)$: $u(\mu^*(w, p)) < 0$, thus the agent rejects the contract.

Subsubcase $w \geq 2\sqrt{p\lambda} - \lambda$: $u(\mu^*(w, p)) \geq 0$, therefore the agent would accept the contract if offered.

Case $p \in (\overline{p}_1, 4\overline{p}_1]$: According to Table 5.1, there is a service capacity that maximizes $u(\mu)$ for $\mu \in (0, \lambda]$ and a service capacity that maximizes $u(\mu)$ for $\mu > \lambda$. Denote the optimal service capacity in $(0, \lambda]$ by $\mu^*_{(0,\lambda]}(w, p)$. From first order condition the optimal service capacity is $\mu^*_{(0,\lambda]}(w, p) = \sqrt{(1-\overline{\eta})p\lambda} - \lambda$ and from (5.3) $u\left(\mu^*_{(0,\lambda]}(w, p)\right) = w - \overline{\eta}p/2 - 2\sqrt{(1-\overline{\eta})p\lambda} + \lambda$. Denote the optimal service capacity for $\mu > \lambda$ by $\mu^*_\lambda(w, p)$. From first order condition $\mu^*_\lambda(w, p) = \sqrt{p\lambda} - \lambda$ and from (5.3) $u\left(\mu^*_\lambda(w, p)\right) = w - 2\sqrt{p\lambda} + \lambda$. The agent has a choice of two service capacities and he installs the one that generates a higher expected utility rate. Note that $u\left(\mu^*_\lambda(w, p)\right) - u\left(\mu^*_{(0,\lambda]}(w, p)\right) = \overline{\eta}p/2 - 2\left(1 - \sqrt{1-\overline{\eta}}\right)\sqrt{p\lambda}$. According to Lemma 5.8, $\overline{p}_1 \geq \overline{p}_2$, therefore according to Lemma 5.7 part (b), $u\left(\mu^*_\lambda(w, p)\right) > u\left(\mu^*_{(0,\lambda]}(w, p)\right)$, the agent's optimal service capacity is $\mu^*(w, p) = \sqrt{p\lambda} - \lambda$ and $u(\mu^*(w, p)) = w - 2\sqrt{p\lambda} + \lambda$. Since $p > \overline{p}_1 > 4\lambda \Rightarrow 2\sqrt{p\lambda} - \lambda > 3\lambda > 0$, therefore we examine the following subcases.

Subcase $w \in \left(0, 2\sqrt{p\lambda} - \lambda\right)$: $u(\mu^*(w, p)) < 0$, therefore the agent rejects the contract.

Subcase $w \geq 2\sqrt{p\lambda} - \lambda$: $u(\mu^*(w, p)) \geq 0$, therefore the agent would accept the contract if offered.

Case $p > 4\overline{p}_1$: According to Table 5.1, the service capacity that maximizes $u(\mu)$ satisfies $\mu > \lambda$. From the first order condition the agent's optimal service capacity is $\mu^*(w, p) = \sqrt{p\lambda} - \lambda$ and $u(\mu^*(w, p)) = w - 2\sqrt{p\lambda} + \lambda$. Since $p > 4\overline{p}_1 > 4\lambda \Rightarrow 2\sqrt{p\lambda} - \lambda > 3\lambda > 0$, therefore we examine the following subcases.

Subcase $w \in \left(0, 2\sqrt{p\lambda} - \lambda\right)$: $u(\mu^*(w, p)) < 0$, thus the agent rejects the contract.

Subcase $w \geq 2\sqrt{p\lambda} - \lambda$: $u(\mu^*(w, p)) \geq 0$, therefore the agent would accept the contract if offered.

This complete the proof for Proposition 5.19 when $\overline{\eta} \in [8/9, 1)$.

Case $\overline{\eta} \in [1, 2)$: Note that $4\lambda > 0 > \overline{p}_1 > 4\overline{p}_1$ and according to Lemma 5.18, $\overline{p}_3 > 4\lambda$. Therefore $\overline{p}_3 > 4\lambda > 0 > \overline{p}_1 > 4\overline{p}_1$. Figure 5.10 depicts the shape of $u(\mu)$ when $\overline{\eta} \in [1, 2)$ and the value of p falls in different ranges. The structure of the proof when $\overline{\eta} \in [1, 2)$ is depicted in Fig. 5.11.

Case $p \in (0, 4\lambda]$: According to Table 5.1, $u(\mu)$ is decreasing with respect to $\mu \geq 0$. Therefore the agent's optimal service capacity is $\mu^*(w, p) = 0$ and from Eq. (5.3) $u(\mu^*(w, p)) = w - (1 - \overline{\eta}/2)p$. Note that $1 - \overline{\eta}/2 > 0$.

Subcase $w \in (0, (1 - \overline{\eta}/2)p)$: $u(\mu^*(w, p)) < 0$, therefore the agent rejects the contract.

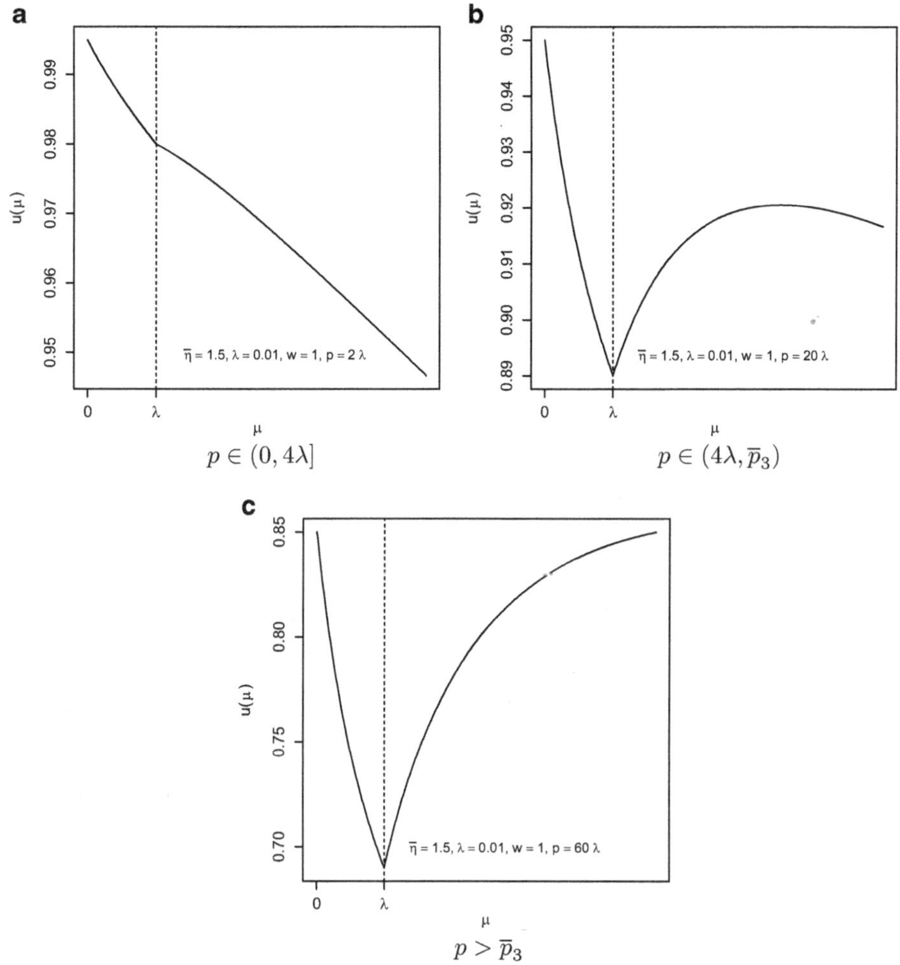

Fig. 5.10 Illustration of the forms of $u(\mu)$ when $\overline{\eta} \in [1, 2)$

Subcase $w \geq (1 - \overline{\eta}/2)p$: $u(\mu^*(w,p)) \geq 0$, thus the agent would accept the contract if offered.

Case $p > 4\lambda$: According to Table 5.1, there is a service capacity that maximizes $u(\mu)$ for $\mu \in [0, \lambda)$ and a service capacity that maximizes $u(\mu)$ for $\mu > \lambda$. Denote the optimal service capacity in $[0, \lambda)$ by $\mu^*_{[0,\lambda)}(w,p)$. Since $u(\mu)$ is decreasing with respect to μ over $[0, \lambda)$, therefore $\mu^*_{[0,\lambda)}(w,p) = 0$ and from (5.3) $u\left(\mu^*_{[0,\lambda)}(w,p)\right) = w - (1 - \overline{\eta}/2)p$. Denote the optimal service capacity for $\mu > \lambda$ by $\mu^*_\lambda(w,p)$. From first order condition $\mu^*_\lambda(w,p) = \sqrt{p\lambda} - \lambda$ and from (5.3) $u\left(\mu^*_\lambda(w,p)\right) = w - 2\sqrt{p\lambda} + \lambda$. The agent has to choose one of the

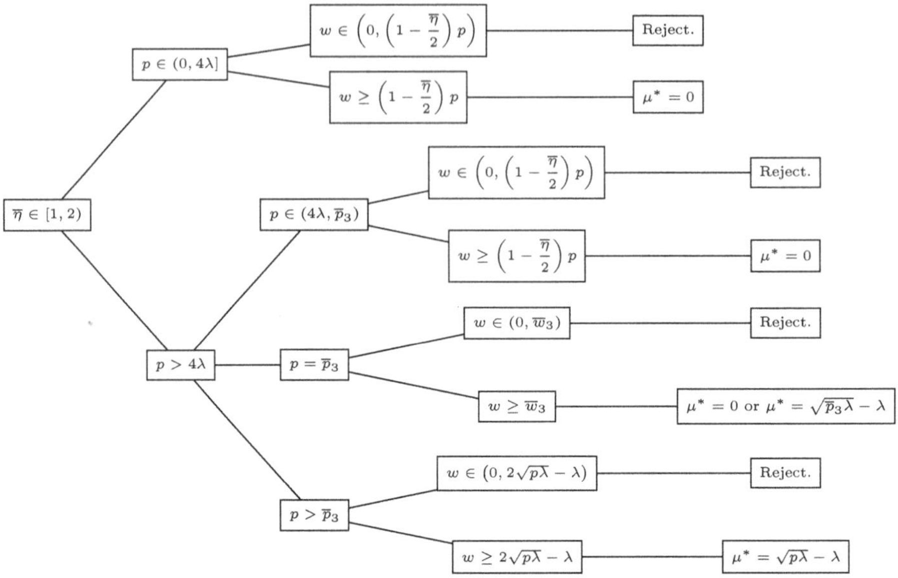

Fig. 5.11 Structure of the proof for Proposition 5.19 when $\overline{\eta} \in [1, 2)$

two service capacities and he installs the one with higher expected utility rate. Note that $u\left(\mu_\lambda^*(w,p)\right) - u\left(\mu_{[0,\lambda)}^*(w,p)\right) = (1 - \overline{\eta}/2)p - 2\sqrt{p\lambda} + \lambda$. According to Lemma 5.5, $4\lambda > 2\lambda/\left(2 + \overline{\eta} + 2\sqrt{2\overline{\eta}}\right)$ and according to Lemma 5.18, $\overline{p}_3 > 4\lambda$, therefore we examine the following subcases.

Subcase $p \in (4\lambda, \overline{p}_3)$: By Lemma 5.4 part (a), $u\left(\mu_{[0,\lambda)}^*(w,p)\right) > u\left(\mu_\lambda^*(w,p)\right)$, therefore the agent's optimal service capacity is $\mu^*(w,p) = 0$ and $u(\mu^*(w,p)) = w - (1 - \overline{\eta}/2)p$. Note that $1 - \overline{\eta}/2 > 0$.

Subsubcase $w \in (0, (1 - \overline{\eta}/2)p)$: $u(\mu^*(w,p)) < 0$, therefore the agent rejects the contract.

Subsubcase $w \geq (1 - \overline{\eta}/2)p$: $u(\mu^*(w,p)) \geq 0$, thus the agent would accept the contract if offered.

Subcase $p = \overline{p}_3$: According to Lemma 5.4 part (c), $u\left(\mu_{[0,\lambda)}^*(w,p)\right) = u\left(\mu_\lambda^*(w,p)\right)$, indicating that installing $\mu_{[0,\lambda)}^*(w,\overline{p}_3)$ or $\mu_\lambda^*(w,\overline{p}_3)$ leads to the same agent's expected utility rate. Therefore the agent is indifferent about installing $\mu^*(w,p) = 0$ or $\mu^*(w,p) = \sqrt{p\lambda} - \lambda$. Still, the capacity value has to lead to admissible solutions (see Proposition 5.20). Recall that by definition of $\overline{w}_3 = (1 - \overline{\eta}/2)\overline{p}_3$ (see (5.24)) and $1 - \overline{\eta}/2 > 0 \Rightarrow \overline{w}_3 > 0$.

Subsubcase $w \in (0, \overline{w}_3)$: $u(\mu^*(w,p)) < 0$, therefore the agent rejects the contract.

Subsubcase $w \geq \overline{w}_3$: $u(\mu^*(w,p)) \geq 0$, thus the agent would accept the contract if offered.

Subcase $p > \overline{p}_3$: From Lemma 5.4 part (b), $u\left(\mu^*_\lambda(w,p)\right) > u\left(\mu^*_{[0,\lambda)}(w,p)\right)$, therefore the agent's optimal service capacity is $\mu^*(w,p) = \sqrt{p\lambda} - \lambda$ and $u(\mu^*(w,p)) = w - 2\sqrt{p\lambda} + \lambda$. Since $p > \overline{p}_3 > 4\lambda \Rightarrow 2\sqrt{p\lambda} - \lambda > 3\lambda > 0$, therefore we examine the following subcases.

Subsubcase $w \in \left(0, 2\sqrt{p\lambda} - \lambda\right)$: $u(\mu^*(w,p)) < 0$, thus the agent rejects the contract.

Subsubcase $w \geq 2\sqrt{p\lambda} - \lambda$: $u(\mu^*(w,p)) \geq 0$, therefore the agent would accept the contract if offered.

This completes the proof for Proposition 5.19 when $\overline{\eta} \in [1,2)$. □

In summary, under the exogenous market conditions such that a contract between the principal and a moderately risk-seeking agent is feasible (see Theorem 5.22 later), only one formula is needed for the agent to compute his optimal service capacity: $\mu^*(w,p) = \sqrt{p\lambda} - \lambda > 0$.

The conditions when a moderately risk-seeking agent accepts the contract can be depicted by the shaded areas in Fig. 5.12, where $\overline{\eta} = 1$. The two shaded areas with different grey scales represent conditions (5.25) and (5.27) under which the agent accepts the contract but responds differently. The lower bound function of the shaded area (denoted by $w_0(p)$) represents the set of offers that give the agent zero expected utility rate. $w_0(p)$ is defined as follows:

$$w_0(p) = \begin{cases} \left(1 - \dfrac{\overline{\eta}}{2}\right) p \text{ when } p \in (0, \overline{p}_3] \\ 2\sqrt{p\lambda} - \lambda \text{ when } p > \overline{p}_3 \end{cases}$$

Since $\lim_{p\to\overline{p}_3^-} w_0(p) = \lim_{p\to\overline{p}_3^+} w_0(p) = \left(\sqrt{2} + \sqrt{\overline{\eta}}\right)\lambda / \left(\sqrt{2} - \sqrt{\overline{\eta}}\right)$, therefore $w_0(p)$ is continuous everywhere over interval $p \in \mathbb{R}_+$. However since $\lim_{p\to\overline{p}_3^-} dw_0(p)/dp = 1 - \overline{\eta}/2 \neq 1 - \sqrt{\overline{\eta}/2} = \lim_{p\to\overline{p}_3^+} dw_0(p)/dp$, therefore $w_0(p)$ is not differentiable at $p = \overline{p}_3$.

5.2.1 *Sensitivity Analysis of a Moderately Risk-Seeking Agent's Optimal Strategy*

Since the principal does not propose a contract that even if accepted will result in zero service capacity, therefore the only viable case is when the agent accepts the contract and installs positive service capacity: $\mu^*(w,p) = \sqrt{p\lambda} - \lambda$. In such a case the agent's optimal strategy is identical to the optimal strategy for risk-neutral agent. According to (5.10) the compensation rate w is bounded below by $2\sqrt{p\lambda} - \lambda = pP(1) + \mu^*(w,p)$, with the term $pP(1)$ representing the expected penalty rate charged by the principal when the optimal capacity is installed. It indicates that the

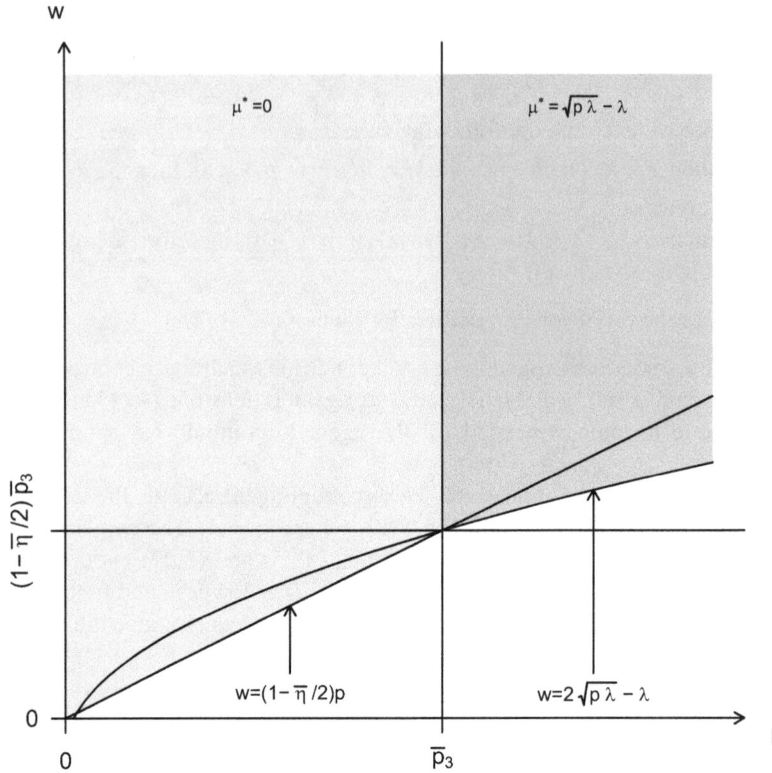

Fig. 5.12 Conditions when a moderately risk-seeking agent accepts the contract with $\overline{\eta} = 1$

agent has to be reimbursed for the expected penalty rate and the cost of service capacity.

The optimal service capacity $\sqrt{p\lambda} - \lambda$ depends on the penalty rate p and the failure rate λ. Its derivatives are $\partial\mu^*/\partial p = \sqrt{\lambda/4p} > 0$ and $\partial\mu^*/\partial\lambda = \sqrt{p/4\lambda} - 1$. These derivatives suggest that given the failure rate, the agent will increase the service capacity when the penalty rate increases. Note that $\sqrt{p\lambda} - \lambda$, as a function of λ, increases when $p/4 > \lambda$. From conditions (5.26) and (5.27) the agent installs service capacity $\sqrt{p\lambda} - \lambda$ when $p \geq \overline{p}_3$, and according to Lemma 5.18 we have $\overline{p}_3 > 4\lambda$. Therefore we have $p > 4\lambda \Rightarrow p/4 > \lambda \Rightarrow \partial\mu^*/\partial\lambda > 0$. Thus, given the penalty rate, the agent will increase the service capacity when the failure rate increases.

The agent's optimal expected utility rate when installing capacity $\mu^*(w, p) = \sqrt{p\lambda} - \lambda$ is $u_A^* \equiv u_A(\mu^*(w, p); w, p) = w - 2\sqrt{p\lambda} + \lambda$, and it depends on w, p and λ only. Note that $\partial u_A^*/\partial w = -1 < 0$, $\partial u_A^*/\partial p = -\sqrt{\lambda/p} < 0$, indicating that the agent's optimal expected utility rate decreases with the compensation rate and the penalty rate. Note that $\partial u_A^*/\partial\lambda = -\sqrt{p/\lambda} + 1$, and from Proposition 5.10

$p \geq \overline{p}_3 > 4\lambda \Rightarrow -\sqrt{p/\lambda} + 1 < 0$, therefore the agent's optimal expected utility rate also decreases with the failure rate.

Summary: Recall that given the set of offers $\{(w,p) : p \in (0,\lambda], w \geq p\}$ a risk-neutral agent would accept the contract, install $\mu^*(w,p) = 0$. When $\left\{(w,p) : p > \lambda, w \geq 2\sqrt{p\lambda} - \lambda\right\}$ he would accept the contract, install $\mu^*(w,p) = \sqrt{p\lambda} - \lambda$ and realize an expected utility rate $u(\mu^*(w,p); w,p) = w - 2\sqrt{p\lambda} + \lambda$. By comparing the optimal capacities of a moderately risk-seeking agent to that of a risk-neutral agent, three conclusions are drawn.

1. The principal has to set a higher p in order to induce a moderately risk-seeking agent to install a positive service capacity versus a risk-neutral agent ($p > \lambda$ for risk-neutral agent, $p > \overline{p}_3 = 2\lambda / \left(\sqrt{2} - \sqrt{\overline{\eta}}\right)^2 > \lambda$ for moderately risk-seeking agent).
2. A moderately risk-seeking agent would install the same positive service capacity as a risk-neutral agent ($\sqrt{p\lambda} - \lambda$).
3. Given w and p, an agent who accepts the contract and subsequently installs a positive service capacity will receive the same expected utility rate when he is moderately risk-seeking ($\overline{\eta} \in [8/9, 2)$) as risk-neutral ($\overline{\eta} = 0$).

5.2.2 Principal's Optimal Strategy

We now proceed to derive the principal's optimal strategy. Anticipating the agent's optimal selection of $\mu^*(w,p)$ the principal chooses w and p to maximize her expected profit rate by solving the optimization problem

$$\max_{w>0,p>0} \Pi_P(w,p;\mu^*(w,p)) = \max_{w>0,p>0} \left\{ \frac{r\mu^*(w,p)}{\lambda + \mu^*(w,p)} - w + \frac{p\lambda}{\lambda + \mu^*(w,p)} \right\} \tag{5.28}$$

where the principal's optimal solution values are $(w^*, p^*) = \text{argmax}_{w>0,p>0} \Pi_P(w,p;\mu^*(w,p))$.

Before we describe the principal's optimal strategy, we reexamine the case when the principal offers $p = \overline{p}_3$ and $w \geq \overline{w}_3$, under which the agent is indifferent about installing two different service capacities. The selected solutions $((w,p),\mu)$ have to be admissible solutions (see Definition 5.1). We state this case formally in Proposition 5.20.

Proposition 5.20. *Suppose a moderately risk-seeking agent and principal. Assume that the principal's offers are constrained to* $\{(w,p) : p = \overline{p}_3, w \geq \overline{w}_3\}$.

(a) *If* $r \in (0, \overline{p}_3]$, *then the principal does not propose a contract.*
(b) *If* $r > \overline{p}_3$, *then the agent installs* $\mu^* = \sqrt{\overline{p}_3\lambda} - \lambda$ *if offered a contract.*

Proof. Note that for $w \geq \overline{w}_3$ we have $\partial \Pi_P(w, \overline{p}_3; \mu)/\partial \mu = (r - \overline{p}_3)\lambda/(\lambda + \mu)^2$. Define $\mu_L \equiv 0$ and $\mu_H \equiv \sqrt{\overline{p}_3 \lambda} - \lambda$ and note that $\mu_H > \mu_L$. If $r \in (0, \overline{p}_3)$, then $\partial \Pi_P/\partial \mu < 0$, therefore $((w, \overline{p}_3), \mu_L) \succeq ((w, \overline{p}_3), \mu_H)$ and the agent would install μ_L if offered a contract. However condition (c) in Definition 5.1 requires that $w \geq \overline{p}_3$, therefore $\Pi_P(w, \overline{p}_3; \mu_L) = -w + \overline{p}_3 \leq 0$ and the principal would not propose a contract. If $r = \overline{p}_3$, then $\partial \Pi_P/\partial \mu = 0$, therefore the agent installs either μ_L or μ_H if offered a contract. However in such case the principal's expected profit rate is $\Pi_P(w, \overline{p}_3; \mu_L) = \Pi_P(w, \overline{p}_3; \mu_H) = -w + \overline{p}_3$, which is non-positive due to condition (c) in Definition 5.1, thus the principal would not propose a contract. If $r > \overline{p}_3$, then $\partial \Pi_P/\partial \mu > 0$ and $((w, \overline{p}_3), \mu_H) \succeq ((w, \overline{p}_3), \mu_L)$. If the principal offers a contract (where the conditions will be discussed in detail in Theorem 5.22 that follows), then by Definition 5.1 only μ_H leads to admissible solutions. □

Lemma 5.21. *Consider* $\max_{x \geq \sqrt{\overline{p}_3}} f(x)$ *where* $f(x) = r + \lambda - \sqrt{\lambda}\,(x + r/x)$ *and denote* $x^* = \operatorname{argmax}_{x \geq \sqrt{\overline{p}_3}} f(x)$. *The solutions to this optimization problem are (see proof in the Appendix):*

(a) $x^* = \sqrt{\overline{p}_3}$ *if* $r \in (0, \overline{p}_3]$.
(b) $x^* = \sqrt{r}$ *if* $r > \overline{p}_3$.

The principal's optimal strategy is described in Theorem 5.22. Recall that Proposition 5.19 describes the agent's optimal response to each pair $(w, p) \in \mathbb{R}^2_+$. Since the principal will not propose a contract that is going to be rejected by a moderately risk-seeking (MRS) agent, therefore Theorem 5.22 only considers pairs $(w, p) \in \mathbb{R}^2_+$ that result in agent's non-negative expected utility rate. Define

$$\begin{aligned} \mathfrak{D}_{(5.25)} &\equiv \{(w,p) \text{ that satisfies (5.25) when } \overline{\eta} \in [8/9, 2)\} \\ \mathfrak{D}_{(5.26)} &\equiv \{(w,p) \text{ that satisfies (5.26) when } \overline{\eta} \in [8/9, 2)\} \\ \mathfrak{D}_{(5.27)} &\equiv \{(w,p) \text{ that satisfies (5.27) when } \overline{\eta} \in [8/9, 2)\} \\ \mathfrak{D}_{\text{MRS}} &\equiv \mathfrak{D}_{(5.25)} \cup \mathfrak{D}_{(5.26)} \cup \mathfrak{D}_{(5.27)} \end{aligned} \tag{5.29}$$

Theorem 5.22. *Given a moderately risk-seeking agent and* $(w, p) \in \mathfrak{D}_{\text{MRS}}$.

(a) *If* $r \in (0, \overline{p}_3]$, *then the principal does not propose a contract.*
(b) *If* $r > \overline{p}_3$, *then the principal's offer and the capacity installed by the agent are*

$$(w^*, p^*) = \left(2\sqrt{r\lambda} - \lambda, r\right) \text{ and } \mu^*(w^*, p^*) = \sqrt{r\lambda} - \lambda \tag{5.30}$$

and the principal's expected profit rate is

$$\Pi_P(w^*, p^*; \mu^*(w^*, p^*)) = r - 2\sqrt{r\lambda} + \lambda \tag{5.31}$$

Proof. The structure of the proof for Theorem 5.22 is depicted in Fig. 5.13.

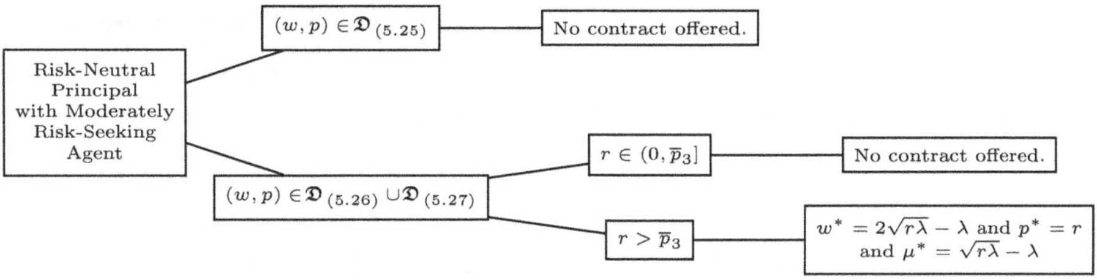

Fig. 5.13 Structure of the proof for Theorem 5.22

Case $(w,p) \in \mathfrak{D}_{(5.25)}$: According to Proposition 5.19 part (a), in case the principal makes an offer, the agent accepts the contract but does not install any service capacity. Since $\partial \Pi_P/\partial w = -1 < 0$, thus we have $w^* = (1 - \overline{\eta}/2)p$ and from (3.3) $\Pi_P(w^*, p; \mu^*(w^*, p)) = -w^* + p = \overline{\eta}p/2 > 0$. However in such case $p > w^* = (1 - \overline{\eta}/2)p$, which violates condition (c) in Definition 5.1, therefore $((w^* = (1 - \overline{\eta}/2)p, p), \mu^* = 0)$ is not an admissible solution and the principal does not propose a contract.

Case $(w,p) \in \mathfrak{D}_{(5.26)} \cup \mathfrak{D}_{(5.27)}$: According to Proposition 5.19 part (c), if $(w,p) \in \mathfrak{D}_{(5.27)}$, then in case the principal makes an offer, the agent accepts the contract and installs $\sqrt{p\lambda} - \lambda$. Since $\partial \Pi_P/\partial w = -1 < 0$, therefore $w^* = 2\sqrt{p\lambda} - \lambda$. According to Propositions 5.19 part (b) and 5.20, if $(w,p) \in \mathfrak{D}_{(5.26)}$ (which implies $p = \overline{p}_3$), then the principal does not propose a contract if $r \in (0, \overline{p}_3]$, or installs $\sqrt{\overline{p}_3\lambda} - \lambda$ in case the principal makes an offer when $r > \overline{p}_3$. Since $\partial \Pi_P/\partial w = -1 < 0$, therefore $w^* = \overline{w}_3$. Denote the principal's expected profit rate when $(w,p) = (\overline{w}_3, \overline{p}_3)$ and $\mu^*(w,p) = 0$ by $\Pi_P^L(\overline{p}_3)$, and denote the principal's expected profit rate when $(w,p) = (\overline{w}_3, \overline{p}_3)$ and $\mu^*(w,p) = \sqrt{\overline{p}_3\lambda} - \lambda$ by $\Pi_P^H(\overline{p}_3)$. By plugging the value of w, p and μ into (3.3):

$$\Pi_P^L(\overline{p}_3) = -\left(1 - \frac{\overline{\eta}}{2}\right)\overline{p}_3 + \overline{p}_3 = \frac{\overline{\eta}\overline{p}_3}{2} \tag{5.32}$$

$$\Pi_P^H(\overline{p}_3) = r + \lambda - \sqrt{\lambda}\left(\sqrt{\overline{p}_3} + \frac{r}{\sqrt{\overline{p}_3}}\right) \tag{5.33}$$

In such case the principal's optimization problem is $\max_{p \geq \overline{p}_3} \Pi_P(w^*, p; \mu^*(w^*, p))$ where:

$$\Pi_P(w^*, p; \mu^*(w^*, p)) = \begin{cases} \max\left\{\Pi_P^L(\overline{p}_3), \Pi_P^H(\overline{p}_3)\right\}, \text{for } p = \overline{p}_3 \\ r + \lambda - \sqrt{\lambda}\left(\sqrt{p} + \dfrac{r}{\sqrt{p}}\right), \text{for } p > \overline{p}_3 \end{cases}$$

Define $x \equiv \sqrt{p}$, the expression $r + \lambda - \sqrt{\lambda}\left(\sqrt{p} + r/\sqrt{p}\right)$ can be restated as $f(x) = r + \lambda - \sqrt{\lambda}\,(x + r/x)$. Maximizing $f(x)$ with respect to $x \geq \sqrt{\overline{p}_3}$ is equivalent to maximizing $r + \lambda - \sqrt{\lambda}\left(\sqrt{p} + r/\sqrt{p}\right)$ with respect to $p \geq \overline{p}_3$ in the sense that

$$\underset{p \geq \overline{p}_3}{\operatorname{argmax}} \left\{ r + \lambda - \sqrt{\lambda}\left(\sqrt{p} + \frac{r}{\sqrt{p}}\right)\right\} = \left(\underset{x \geq \sqrt{\overline{p}_3}}{\operatorname{argmax}} f(x)\right)^2$$

Therefore we examine the following subcases.

Subcase $r \in (0, \overline{p}_3]$: According to Lemma 5.21 part (a), $p^* = \overline{p}_3$ and according to Proposition 5.20 part (a) the principal does not propose a contract.

Subcase $r > \overline{p}_3$: According to Lemma 5.21 part (b), $p^* = r$ and the principal's expected profit rate is $\Pi_P(w^*, p^*; \mu^*(w^*, p^*)) = r - 2\sqrt{r\lambda} + \lambda > \Pi_P^H(\overline{p}_3) > \Pi_P^L(\overline{p}_3) = \overline{\eta}\overline{p}_3/2 > 0$. Thus the principal proposes a contract with $w^* = 2\sqrt{r\lambda} - \lambda$ and $p^* = r$ that induces the agent to install $\mu^*(w^*, p^*) = \sqrt{r\lambda} - \lambda$.

To summarize, if $r \in (0, \overline{p}_3]$, then the principal does not propose a contract. If $r > \overline{p}_3$, then the principal offers $(w^*, p^*) = \left(2\sqrt{r\lambda} - \lambda, r\right)$ and the agent installs capacity $\mu^*(w^*, p^*) = \sqrt{r\lambda} - \lambda$. □

Theorem 5.22 indicates that the existence of a contract for moderately risk-seeking agent is determined exogenously by the r, λ, and $\overline{\eta}$.

5.3 Optimal Strategies for the Strongly Risk-Seeking Agent

We start by deriving the strongly risk-seeking agent's optimal strategy. The agent's optimization problem is defined in (5.3).

First a technical lemma (see proof in the Appendix).

Lemma 5.23. *Let $\overline{\eta} > 2$ and $\lambda > 0$.*

(a) *If* $\dfrac{2\lambda}{2 + \overline{\eta} + 2\sqrt{2\overline{\eta}}} > p > 0$, *then* $\left(1 - \dfrac{\overline{\eta}}{2}\right)p - 2\sqrt{p\lambda} + \lambda > 0.$

(b) *If* $p > \dfrac{2\lambda}{2 + \overline{\eta} + 2\sqrt{2\overline{\eta}}}$, *then* $0 > \left(1 - \dfrac{\overline{\eta}}{2}\right)p - 2\sqrt{p\lambda} + \lambda.$

(c) *If* $p = \dfrac{2\lambda}{2 + \overline{\eta} + 2\sqrt{2\overline{\eta}}}$, *then* $\left(1 - \dfrac{\overline{\eta}}{2}\right)p - 2\sqrt{p\lambda} + \lambda = 0.$

We describe a strongly risk-seeking agent's optimal response to any possible offered contract $(w, p) \in \mathbb{R}_+^2$ in Proposition 5.24.

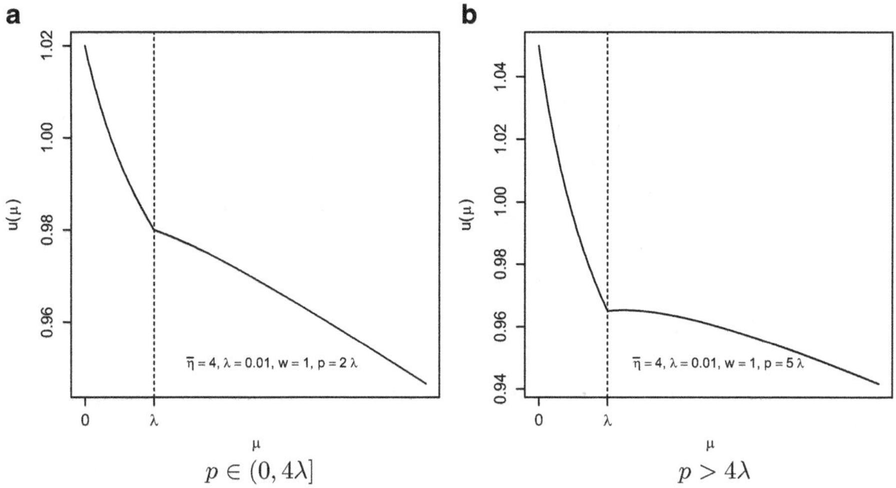

Fig. 5.14 Illustration of the forms of $u(\mu)$ when $\overline{\eta} \geq 2$

Fig. 5.15 Structure of the proof for Proposition 5.24 when $\overline{\eta} \geq 2$

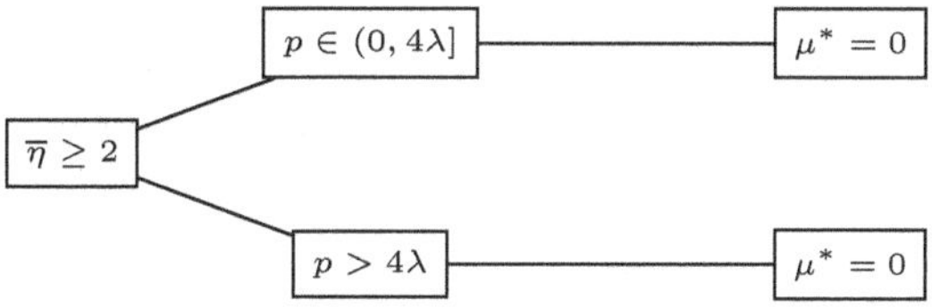

Proposition 5.24. *Consider a strongly risk-seeking agent* ($\overline{\eta} \geq 2$). $\forall\ w > 0$ *and* $p > 0$, *the agent accepts the contract and installs* $\mu^*(w,p) = 0$ *with resulting expected utility rate* $u_A(\mu^*(w,p);w,p) = w - (1 - \overline{\eta}/2)p > 0$.

Proof. Figure 5.14 shows the shape of $u(\mu)$ when $\overline{\eta} \geq 2$ and the value of p falls in different ranges. The structure of the proof when $\overline{\eta} \geq 2$ is depicted in Fig. 5.15.

Case $p \in (0, 4\lambda]$: According to Table 5.1, $u(\mu)$ is decreasing with respect to $\mu \geq 0$. Therefore the agent's optimal service capacity is $\mu^*(w,p) = 0$ and from (5.3) $u(\mu^*(w,p)) = w - (1 - \overline{\eta}/2)p$. Note that $1 - \overline{\eta}/2 \leq 0$, therefore $\forall\ w > 0$, $u(\mu^*(w,p)) > 0$ and the agent would accept the contract if offered.

Case $p > 4\lambda$: According to Table 5.1, there is a service capacity that maximizes $u(\mu)$ for $\mu \in [0, \lambda)$ and a service capacity that maximizes $u(\mu)$ for $\mu > \lambda$. Denote the optimal service capacity in $[0, \lambda)$ by $\mu^*_{[0,\lambda)}(w,p)$. Since $u(\mu)$ is decreasing with respect to μ over $[0, \lambda)$, therefore $\mu^*_{[0,\lambda)}(w,p) = 0$ and from (5.3) $u\left(\mu^*_{[0,\lambda)}(w,p)\right) = w - (1 - \overline{\eta}/2)p$. Denote the optimal service capacity for $\mu > \lambda$ by $\mu^*_{\lambda}(w,p)$. From first order condition $\mu^*_{\lambda}(w,p) = \sqrt{p\lambda} - \lambda$ and from (5.3) $u\left(\mu^*_{\lambda}(w,p)\right) = w - 2\sqrt{p\lambda} + \lambda$. The agent has to choose one of the two service capacities and he installs the one with higher expected utility rate. Note that $u\left(\mu_{\lambda}^{*}(w,p)\right) - u\left(\mu_{[0,\lambda)}^{*}(w,p)\right) = (1-\overline{\eta}/2)p - 2\sqrt{p\lambda} + \lambda$. If $\overline{\eta} = 2$, then $u\left(\mu_{\lambda}^{*}(w,p)\right) - u\left(\mu_{[0,\lambda)}^{*}(w,p)\right) = -2\sqrt{p\lambda} + \lambda$, and since $p > 4\lambda \Leftrightarrow 2\sqrt{p\lambda} > 4\lambda \Leftrightarrow 0 > -3\lambda > -2\sqrt{p\lambda}+\lambda$, we have $u\left(\mu_{[0,\lambda)}^{*}(w,p)\right) > u\left(\mu_{\lambda}^{*}(w,p)\right)$. If $\overline{\eta} > 2$, then according to Lemmas 5.5 and 5.23 part (b), $p > 4\lambda > 2\lambda/\left(2+\overline{\eta}+2\sqrt{2\overline{\eta}}\right) \Rightarrow u\left(\mu_{[0,\lambda)}^{*}(w,p)\right) > u\left(\mu_{\lambda}^{*}(w,p)\right)$. Thus the agent's optimal service capacity is $\mu^*(w,p) = 0$ and $u(\mu^*(w,p)) = w - (1-\overline{\eta}/2)p$. Note that $1-\overline{\eta}/2 \leq 0$, therefore $\forall\, w > 0$, $u(\mu^*(w,p)) > 0$ and the agent would accept the contract if offered. □

Proposition 5.24 indicates that a strongly risk-seeking agent does not commit any capacity, therefore the principal does not propose any contract, which we state formally in Theorem 5.25.

Theorem 5.25. *A principal never offers a contract to a strongly risk-seeking agent.*

Proof. According to Proposition 5.24, the agent accepts the contract but does not install any service capacity for all $(w,p) \in \mathbb{R}_{+}^{2}$. In such case the principal's expected profit rate is $\Pi_P(w,p;\mu^*(w,p)) = -w + p$. Since condition (c) of Definition 5.1 requires that $w \geq p$, therefore $\Pi_P(w,p;\mu^*(w,p)) \leq 0$ and the principal does not propose a contract to a strongly risk-seeking agent! □

5.4 Risk-Seeking Agent: A Summary

Recall the definition of $\overline{p}_2$, $\overline{r}_2$ and $\overline{p}_3$ from (5.5), (5.12) and (5.23). The conditions when a principal makes contract offers to a risk-seeking agent is depicted by the shaded areas in Fig. 5.16. The horizontal axis represents the agent's risk coefficient $\overline{\eta}$, and the vertical axis represents the revenue rate generated by the principal's equipment unit, which is exogenously determined by the market. The principal makes different offers to the agent when $(r,\overline{\eta})$ is in the three shaded areas with different gray scales. We define

$$\overline{p}_{23} \equiv \begin{cases} \overline{p}_2 \text{ for } \overline{\eta} \in (0, 8/9) \\ \overline{p}_3 \text{ for } \overline{\eta} \in [8/9, 2) \end{cases}$$

Since $\lim_{\overline{\eta}\to(8/9)-}\overline{p}_{23} = \lim_{\overline{\eta}\to(8/9)+}\overline{p}_{23} = 9\lambda$ and $\lim_{\overline{\eta}\to(8/9)-}\partial\overline{p}_{23}/\partial\overline{\eta} = \lim_{\overline{\eta}\to(8/9)+}\partial\overline{p}_{23}/\partial\overline{\eta} = 81\lambda/4$, and note that $\lim_{\overline{\eta}\to 2-}\overline{p}_{23} = \lim_{\overline{\eta}\to 2-}\overline{p}_3 = +\infty$, therefore $\overline{p}_{23}$ is continuous and differentiable everywhere over $(0,2)$. In Fig. 5.16 we only describe the conditions of a risk-neutral principal making offers to a weakly and moderately risk-seeking agent ($\overline{\eta} \in (0,2)$), because the principal never makes a contract offer to a strongly risk-seeking agent ($\overline{\eta} \geq 2$).

The revenue rate parameter r is determined exogenously by the market, and we assume that the principal is only interested in operating the equipment when the

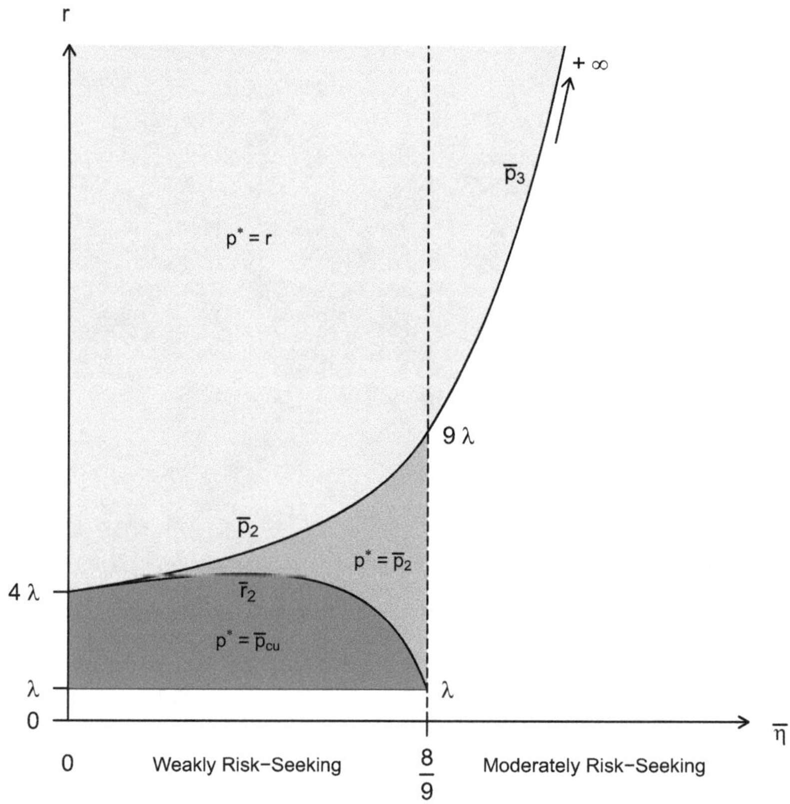

Fig. 5.16 Conditions when a principal makes contract offers to a risk-seeking agent

revenue rate is sufficiently high, specifically $r > \overline{p}_{23}$. In such case weakly and moderately risk-seeking agents would behave exactly the same as a risk-neutral agent, and a strongly risk-seeking agent will never be offered a contract.

Chapter 6
Summary

In this paper we examine a basic principal-agent arrangement for contracting an exclusive equipment repair service supplier. The system setting consists of one principal, one agent, and one revenue generating unit that breaks down from time to time and needs to be repaired when a failure occurs. Our assumptions are that the risk-neutral principal maximizes her expected profit rate given market driven revenue rate r collected during the unit's uptime, the unit's failure rate λ, and the agent's risk attitude η. We consider different agent types – risk neutral, weakly risk-averse, strongly risk-averse, weakly risk-seeking, moderate risk-seeking, and strongly risk-seeking. As is common in a principal-agent context the principal cannot contract directly for the agent's service capacity μ. The nature of the principal-agent contract is that the principal supports the agent at a compensation rate $w > 0$ but imposes on the agent a penalty rate $p > 0$ during the time the unit is down. We note that the nature of the contract does not change if the w is paid to the agent only during the unit's uptime. In fact, the two contract versions are equivalent (see Observation 3.1).

The main contribution of this paper is in the complete analysis of the contractual details that have to be addressed in the agreement between the unit's owner and the supplier of repair services. Our pedestrian assumptions are that the failure rate of the equipment unit is a constant λ, the repair time duration has an exponential distribution with a constant repair rate μ. Furthermore, we do not restrict the contract to a specific period of time, rather the contract can be for undetermined time. With the assumption that both the principal and the agent are infinitely rational the surprising outcome is that calculating the optimal strategies for the two parties in all circumstances can be accomplished with an aid of small number of formulas – 7 sets in total. That is, given exogenously determined values of market driven revenue rate, equipment's failure rate, repair capacity marginal cost, and the type of a repair agent, it is straight forward to calculate principal's optimal contract offer if one exists, together with agent's optimal service capacity decision. An optimal

S. Zeng, M. Dror, *Formulating Principal-Agent Service Contracts for a Revenue Generating Unit*, SpringerBriefs in Operations Management,
DOI 10.1007/978-3-319-18672-6_6

contract consists of compensation rate w and penalty rate p, both determined by the principal, and the capacity value of μ determined by the agent.

Our analysis of the above principal-agent cooperation is divided into three main parts based on agent's type starting with risk-neutral agent. The second part examines the case of a contracting a risk-averse agent followed by the analysis of a contract given a risk-seeking agent. To our knowledge analysis of principal-agent with risk-seeking agent has not received much coverage in the literature.

As for the analysis of principal-agent construct given a risk-neutral agent, for the entire range of exogenous parameters' values, it can be summarized for the principal by one set of formulas calculating optimal compensation rate $w^* = 2\sqrt{r\lambda} - \lambda$ and optimal penalty rate $p^* = r$. The agent's optimal capacity rate formula is $\mu^*(w^*, p^*) = \sqrt{r\lambda} - \lambda$. We note that this case has the property that without checking if the given market conditions guarantee the existence of a contract, by calculating principal's optimal contract terms w^* and p^* and agent's optimal capacity value $\mu^*(w^*, p^*)$, we simultaneously verify contract existence if the resulting $\mu^*(w^*, p^*)$ is positive. If the optimal capacity value is zero or negative, then it means that the given market conditions do not support a service contract. It also important to note that, for our principal-agent given a risk-neutral agent, if an optimal contract is feasible then it is also efficient.

When considering a risk-averse agent the first task is to decide on the appropriate mathematical expression that captures the agent's disutility with regard to his revenue dispersion. After examining risk premium expressions in the literature we opted for a new risk expression not yet seen in the literature. We express agent's disutility as $\eta p(1/2 - |1/2 - \lambda/(\lambda + \mu)|)$. This measure of agent's utility value due to his revenue fluctuation is introduced and discussed in Chap. 4. The main points are that the risk expression acts like standard deviation and is unit-wise compatible with other terms of agent's utility. In high revenue industry, if the principal contracts with a risk-averse agent with the risk disutility measured by the dispersion of the agent's revenue stream, then agent's risk-aversion reduces the principal's optimal penalty rate and leads to deterioration of the equipment unit's performance. Furthermore, with risk-averse agent the principal is strictly worse off in relation to risk-neutral agent and the social welfare is reduced as the agent's risk-aversion increases.

We divided risk-averse agents into two types based on risk intensity parameter η. That is, for $\eta \in (0, 4/5)$ we refer to the agent as *weakly risk-averse* (Sect. 4.1) and for $\eta \geq 4/5$ we refer to the agent as *strongly risk-averse* (Sect. 4.2). A weakly risk-averse agent has only two formulas to consider: (i) $\mu^*(w, p) = \sqrt{(1-\eta)p\lambda} - \lambda$ or (ii) $\mu^*(w, p) = \sqrt{(1+\eta)p\lambda} - \lambda$. Only one formula, the same as (ii), is sufficient given a strongly risk-averse agent. Formula (i) exists only for WRA agent because when the penalty rate is low, the savings from reducing the service capacity is more prominent than the increase in the penalty charge, providing an incentive for the agent to reduce the optimal service capacity, which deteriorates the performance of the principal's equipment unit. When the penalty rate becomes high, WRA agent increases his service capacity to reduce the penalty charge, which results in formula (ii).

For a risk-seeking agent we adopt a risk premium expression that reflects the expected amount at stake instead of the dispersion of his revenue stream. Our new risk premium expression is consistent with the theoretical developments and empirical evidences regarding the properties of risk in recent literature. We express agent's risk premium as $-\eta p\,(\lambda/(\lambda+\mu)-1/2)_+$, which is unit-wise compatible with other terms of agent's utility (see Chap. 5). If the principal contracts with a risk-seeking agent with low penalty rate, then the agent's risk-seeking deteriorates the performance of the principal's equipment unit. If the principal contract with a risk-seeking agent with high penalty rate, then she can achieve the same equipment performance and contract efficiency as with a risk-neutral agent. However a principal never contracts with a strongly risk-seeking agent.

We categorize risk-seeking agents into three types based on $\overline{\eta}$ – risk intensity parameter. That is, for $\overline{\eta} \in (0, 8/9)$ we refer to the agent as *weakly risk-seeking* (Sect. 5.1), for $\overline{\eta} \in [8/9, 2)$ we refer to the agent as *moderately risk-seeking* (Sect. 5.2) and the agent as *strongly risk-seeking* (Sect. 5.3). A weakly risk-seeking agent has only two formulas to consider: (i) $\mu^*(w,p) = \sqrt{(1-\overline{\eta})p\lambda} - \lambda$ or (ii) $\mu^*(w,p) = \sqrt{p\lambda} - \lambda$. Only one formula, the same as (ii), is sufficient given a moderately risk-seeking agent. A strongly risk-seeking agent never commits any service capacity. Formula (i) exists only for WRS agent because when the penalty rate is low, the risk premium covers the penalty charge thus provides an incentive for the agent to reduce the optimal service capacity compared to risk-neutral. When the penalty rate increases, WRS agent increases his service capacity to reduce the penalty charge that cannot be covered by risk premium, which results in formula (ii).

6.1 Interpreting Table 6.1

Table 6.1 summarizes the formulas for calculating the principal's optimal contract terms and the agent's optimal service capacity when a contract is supported by exogenous market and industry conditions. Mutually exclusive exogenous conditions that support a contract are listed in the column labeled "Exogenous Condition", and the formulas of the principal's optimal contract terms and the agent's optimal capacities are listed in the column labeled "Principal's Formula" and "Agent's Formula" respectively.

If a set of specific market and industry values are observed, namely the value of the agent's risk coefficient η (or $\overline{\eta}$), the revenue rate r, and the failure rate λ, then these values can be validated against the exogenous conditions listed in the table. If the set of values satisfies a certain condition, then the principal's formula and the agent's formula corresponding to that condition can be used to calculate the optimal contract terms and the optimal capacity. No contract is supported if the set of values does not satisfy any condition listed in the table.

To verify that whether the observed values of η, r, and λ satisfy a certain condition, one has to calculate the values that separate the range of r into different

Table 6.1 Summary of the optimal principal-agent contract formulas under exogenous conditions

Exogenous Condition			
Agent's Type	Revenue	Principal's Formula	Agent's Formula
$\eta = 0$ (RN)	$r > \lambda$	$(w^*, p^*) = \left(2\sqrt{r\lambda} - \lambda, r\right)$	$\mu^*(w^*, p^*) = \sqrt{r\lambda} - \lambda$
$\eta \in \left(0, \frac{4}{5}\right)$ (WRA)	$r \in (p_2, p_3]$	$(w^*, p^*) = \left(\eta p_{cu} + 2\sqrt{(1-\eta)p_{cu}\lambda} - \lambda, p_{cu}\right)$	$\mu^*(w^*, p^*) = \sqrt{(1-\eta)p_{cu}\lambda} - \lambda$
	$r \in (p_3, r_2)$	$(w^*, p^*) = \left(\eta p_{cu} + 2\sqrt{(1-\eta)p_{cu}\lambda} - \lambda, p_{cu}\right)$	$\mu^*(w^*, p^*) = \sqrt{(1-\eta)p_{cu}\lambda} - \lambda$
		$(w^*, p^*) = (w_3, p_3)$	$\mu^*(w^*, p^*) = \sqrt{(1+\eta)p_3\lambda} - \lambda$
	$r \in [r_2, r_3]$	$(w^*, p^*) = (w_3, p_3)$	$\mu^*(w^*, p^*) = \sqrt{(1+\eta)p_3\lambda} - \lambda$
	$r > r_3$	$(w^*, p^*) = \left(2\sqrt{\frac{(1+\eta)r\lambda}{1+2\eta}} - \lambda, \frac{r}{1+2\eta}\right)$	$\mu^*(w^*, p^*) = \sqrt{\frac{(1+\eta)r\lambda}{1+2\eta}} - \lambda$
$\eta \geq \frac{4}{5}$ (SRA)	$r \in (p_4, r_4]$	$(w^*, p^*) = (w_4, p_4)$	$\mu^*(w^*, p^*) = \sqrt{(1+\eta)p_4\lambda} - \lambda$
	$r > r_4$	$(w^*, p^*) = \left(2\sqrt{\frac{(1+\eta)r\lambda}{1+2\eta}} - \lambda, \frac{r}{1+2\eta}\right)$	$\mu^*(w^*, p^*) = \sqrt{\frac{(1+\eta)r\lambda}{1+2\eta}} - \lambda$
$\overline{\eta} \in \left(0, \frac{8}{9}\right)$ (WRS)	$r \in (\lambda, \overline{r}_2)$	$(w^*, p^*) = \left(\frac{\overline{\eta} \overline{p}_{cu}}{2} + 2\sqrt{(1-\overline{\eta})\overline{p}_{cu}\lambda} - \lambda, \overline{p}_{cu}\right)$	$\mu^*(w^*, p^*) = \sqrt{(1-\overline{\eta})\overline{p}_{cu}\lambda} - \lambda$
	$r \in [\overline{r}_2, \overline{p}_2]$	$(w^*, p^*) = (\overline{w}_2, \overline{p}_2)$	$\mu^*(w^*, p^*) = \sqrt{(1-\overline{\eta})\overline{p}_2\lambda} - \lambda$
	$r > \overline{p}_2$	$(w^*, p^*) = \left(2\sqrt{r\lambda} - \lambda, r\right)$	$\mu^*(w^*, p^*) = \sqrt{r\lambda} - \lambda$
$\overline{\eta} \in \left[\frac{8}{9}, 2\right)$ (MRS)	$r > \overline{p}_3$	$(w^*, p^*) = \left(2\sqrt{r\lambda} - \lambda, r\right)$	$\mu^*(w^*, p^*) = \sqrt{r\lambda} - \lambda$

intervals, including p_2, p_3, r_2, r_3, p_4, r_4, $\overline{r}_2$, $\overline{p}_2$, and $\overline{p}_3$. Recall that p_2 and p_3 are defined in (4.5), r_2 and r_3 are defined in (4.12), p_4 and w_4 are defined in (4.23), r_4 is defined in (4.28), $\overline{p}_2$ is defined in (5.5), $\overline{r}_2$ is defined in (5.12), and $\overline{p}_3$ is defined in (5.23). Furthermore, to calculate the principal's optimal contract terms and the agent's optimal capacity, one may need to calculate the values of p_{cu}, w_3, $\overline{p}_{cu}$, and $\overline{w}_2$. Recall that p_{cu} can be calculated using (4.13), w_3 is defined in (4.6), $\overline{p}_{cu}$ can be calculated using (5.13), and $\overline{w}_2$ is defined in (5.6).

Specifically, note that when the revenue rate $r \in (p_3, r_2)$, there are two sets of formulas listed in the table to calculate the principal's optimal contract terms and the agent's optimal capacity:

$$(w^*, p^*) = \left(\eta p_{cu} + 2\sqrt{(1-\eta)p_{cu}\lambda} - \lambda, p_{cu}\right), \mu^*(w^*, p^*) = \sqrt{(1-\eta)p_{cu}\lambda} - \lambda$$

$$(w^*, p^*) = (w_3, p_3), \mu^*(w^*, p^*) = \sqrt{(1+\eta)p_3\lambda} - \lambda$$

According to Proposition 4.20 it is difficult to identify the principal's optimal offer when $r \in (p_3, r_2)$ due to the difficulty of computing p_{cu} (see Eq. (4.13)). However, given the value of η, r, and λ, the principal's expected profit rate of both offers can be calculated (see the formulas for calculating the principal's expected profit rate in Proposition 4.20), and the offer with higher expected profit rate should be selected by the principal.

In summary, this paper provides a small set of formulas that exhaustively covers the computing of Pareto optimal principal-agent contract offer and corresponding service capacity for any values of market and industry parameters.

Acknowledgment

We would like to thank Prof. Moshe Shaked for his advice and the many discussions about how to present the mathematical findings of this research. All the remaining errors and misstatements are the authors' alone.

S. Zeng, M. Dror, *Formulating Principal-Agent Service Contracts for a Revenue Generating Unit*, SpringerBriefs in Operations Management,
DOI 10.1007/978-3-319-18672-6

Appendix

Proof of Lemma 4.1. Let $1 > \eta > 0$ and $\lambda > 0$, then $\eta p + 2\sqrt{(1-\eta)p\lambda} - \lambda$ increases with respect to $p > 0$. Further more, if $p = \lambda/(1-\eta)$, then $\eta p + 2\sqrt{(1-\eta)p\lambda} - \lambda = \lambda/(1-\eta)$. Therefore if $p \geq \lambda/(1-\eta)$, then $\eta p + 2\sqrt{(1-\eta)p\lambda} - \lambda \geq \lambda/(1-\eta) > 0$. On the other hand if $p \geq \lambda/(1-\eta)$, then $p - \left(\eta p + 2\sqrt{(1-\eta)p\lambda} - \lambda\right) = \left(\sqrt{(1-\eta)p} - \sqrt{\lambda}\right)^2 \geq 0$, therefore $p \geq \eta p + 2\sqrt{(1-\eta)p\lambda} - \lambda$. □

Proof of Lemma 4.2. Let $1 > \eta > 0$ and $\lambda > 0$. Define $x \equiv \sqrt{p}$ and $a \equiv \sqrt{\lambda}$ and restate the expression $\eta p - 2\left(\sqrt{1+\eta} - \sqrt{1-\eta}\right)\sqrt{p\lambda}$ as $\eta x^2 - 2\left(\sqrt{1+\eta} - \sqrt{1-\eta}\right)ax$ with $x > 0$ and $a > 0$. The solutions to the quadratic equation $\eta x^2 - 2\left(\sqrt{1+\eta} - \sqrt{1-\eta}\right)ax = 0$ for x are 0 and $2\left(\sqrt{1+\eta} - \sqrt{1-\eta}\right)a/\eta$. Therefore if $x > 2\left(\sqrt{1+\eta} - \sqrt{1-\eta}\right)a/\eta$, or equivalently, $x^2 > 8\left(1 - \sqrt{1-\eta^2}\right)a^2/\eta^2$, then $\eta x^2 - 2\left(\sqrt{1+\eta} - \sqrt{1-\eta}\right)ax > 0$. Replacing x by $\sqrt{p}$ and a by $\sqrt{\lambda}$ we obtain (a). The proofs for (b) and (c) are similar. □

Proof of Lemma 4.3. Let $1 > \eta > 0$ and $\lambda > 0$, then we have

$$\begin{aligned}
\left(\sqrt{1+\eta} - \sqrt{1-\eta}\right)^2 > 0 &\Leftrightarrow 1 - \sqrt{1+\eta}\sqrt{1-\eta} > 0 \\
&\Leftrightarrow 1 + \eta - \sqrt{1+\eta}\sqrt{1-\eta} > \eta \\
&\Leftrightarrow (1+\eta)\left(\sqrt{1+\eta} - \sqrt{1-\eta}\right)^2 > \eta^2 \\
&\Leftrightarrow \left(\sqrt{1+\eta} - \sqrt{1-\eta}\right)^2/\eta^2 > 1/(1+\eta) \\
&\Leftrightarrow 8\left(1 - \sqrt{1-\eta^2}\right)\lambda/\eta^2 > 4\lambda/(1+\eta)
\end{aligned}$$

S. Zeng, M. Dror, *Formulating Principal-Agent Service Contracts for a Revenue Generating Unit*, SpringerBriefs in Operations Management,
DOI 10.1007/978-3-319-18672-6

Also we have

$$\begin{aligned}
\left(\sqrt{1+\eta}-\sqrt{1-\eta}\right)^2>0 &\Leftrightarrow 0>\sqrt{1+\eta}\sqrt{1-\eta}-1\\
&\Leftrightarrow \eta>\sqrt{1+\eta}\sqrt{1-\eta}-(1-\eta)\\
&\Leftrightarrow \eta^2>(1-\eta)\left(\sqrt{1+\eta}-\sqrt{1-\eta}\right)^2\\
&\Leftrightarrow 1/(1-\eta)>\left(\sqrt{1+\eta}-\sqrt{1-\eta}\right)^2/\eta^2\\
&\Leftrightarrow 4\lambda/(1-\eta)>8\left(1-\sqrt{1-\eta^2}\right)\lambda/\eta^2
\end{aligned}$$

□

Proof of Lemma 4.4. Let $\eta>0$ and $\lambda>0$. If $p>4\lambda/(1+\eta)$, then $2\sqrt{(1+\eta)p\lambda}-\lambda>4\lambda-\lambda=3\lambda>0$. □

Proof of Lemma 4.5. Let $\eta>0$ and $\lambda>0$. Define $x\equiv\sqrt{p}$ and $a\equiv\sqrt{\lambda}$ and restate expression $p-2\sqrt{(1+\eta)p\lambda}+\lambda$ as $x^2-2ax\sqrt{1+\eta}+a^2$ where $x>0$ and $a>0$. The solutions to the quadratic equation $x^2-2ax\sqrt{1+\eta}+a^2=0$ for x are $\left(\sqrt{1+\eta}-\sqrt{\eta}\right)a$ and $\left(\sqrt{1+\eta}+\sqrt{\eta}\right)a$. Therefore if $\left(\sqrt{1+\eta}+\sqrt{\eta}\right)a>x>\left(\sqrt{1+\eta}-\sqrt{\eta}\right)a$, or equivalently, $\left(1+2\eta+2\sqrt{\eta(1+\eta)}\right)a^2>x^2>\left(1+2\eta-2\sqrt{\eta(1+\eta)}\right)a^2$, then $0>x^2-2ax\sqrt{1+\eta}+a^2$. Replacing x by $\sqrt{p}$ and a by $\sqrt{\lambda}$ we obtain (a). The proofs for (b) and (c) are similar. □

Proof of Lemma 4.6. Let $\eta>0$ and $\lambda>0$. Note that

$$\begin{aligned}
\sqrt{1+\eta}+\sqrt{\eta}>\sqrt{1+\eta}\Rightarrow 2/\sqrt{1+\eta}>1/\sqrt{1+\eta}&>1/\left(\sqrt{1+\eta}+\sqrt{\eta}\right)\\
&\Leftrightarrow 2/(1+\eta)>\sqrt{1+\eta}-\sqrt{\eta}\\
&\Leftrightarrow 4/(1+\eta)>\left(\sqrt{1+\eta}-\sqrt{\eta}\right)^2\\
&\Leftrightarrow 4\lambda/(1+\eta)>\left(1+2\eta-2\sqrt{\eta(1+\eta)}\right)\lambda
\end{aligned}$$

□

Proof of Lemma 4.7. Let $4/5>\eta>0$ and $\lambda>0$, then we have

$$\begin{aligned}
4\eta-5\eta^2>0 &\Leftrightarrow 2\sqrt{\eta(1-\eta)}>\eta\\
&\Leftrightarrow 1+2\sqrt{\eta(1-\eta)}>1+\eta\\
&\Leftrightarrow \left(\sqrt{1-\eta}+\sqrt{\eta}\right)^2>\left(\sqrt{1+\eta}\right)^2
\end{aligned}$$

$$\Leftrightarrow \sqrt{1-\eta}+\sqrt{\eta} > \sqrt{1+\eta}$$
$$\Leftrightarrow \sqrt{1-\eta} > \sqrt{1+\eta}-\sqrt{\eta}$$
$$\Leftrightarrow \sqrt{1-\eta}\left(\sqrt{1+\eta}+\sqrt{\eta}\right) > 1$$
$$\Leftrightarrow \left(1+2\eta+2\sqrt{\eta(1+\eta)}\right)\lambda > \lambda/(1-\eta)$$

and we obtain (a). The proofs for (b) and (c) are similar. □

Proof of Lemma 4.8. Let $4/5 > \eta > 0$ and $\lambda > 0$, then we have

$$9(1-\eta) > 1+\eta \Leftrightarrow 3\sqrt{1-\eta} > \sqrt{1+\eta}$$
$$\Leftrightarrow 4\sqrt{1-\eta} > \sqrt{1+\eta}+\sqrt{1-\eta}$$
$$\Leftrightarrow 2\sqrt{1-\eta}\left(\sqrt{1+\eta}-\sqrt{1-\eta}\right) > \eta$$
$$\Leftrightarrow 4(1-\eta)\left(\sqrt{1+\eta}-\sqrt{1-\eta}\right)^2 > \eta^2$$
$$\Leftrightarrow 8\left(1-\sqrt{1-\eta^2}\right)\lambda/\eta^2 > \lambda/(1-\eta)$$

and we obtain (a). The proofs for (b) and (c) are similar. □

Proof of Lemma 4.13. Let $4/5 > \eta > 0$ and $\lambda > 0$. First we prove (a) and (b) together. According to Lemma 4.8 part (a), $p_3 > p_2$, therefore $p_3 = \eta p_3 + (1-\eta)p_3 > \eta p_3 + (1-\eta)\sqrt{p_2 p_3} > \eta p_2 + (1-\eta)p_2 = p_2$. Next we prove (c). According to Lemma 4.8 part (a), $p_3 > p_2$, therefore $2\eta\left(\sqrt{p_3}-\sqrt{p_2}\right)/\sqrt{p_2} > 0 \Rightarrow \left(1+2\eta\left(\sqrt{p_3}-\sqrt{p_2}\right)/\sqrt{p_2}\right)p_3 > p_3$. Finally we prove (d). According to Lemma 4.3, we have $4p_2 > p_3$, therefore $2\sqrt{p_2} > \sqrt{p_3} \Leftrightarrow \sqrt{p_2} > \sqrt{p_3}-\sqrt{p_2} \Leftrightarrow 1 > \left(\sqrt{p_3}-\sqrt{p_2}\right)/\sqrt{p_2}$. Therefore $(1+2\eta)p_3 > \left(1+2\eta\left(\sqrt{p_3}-\sqrt{p_2}\right)/\sqrt{p_2}\right)p_3$. □

Proof of Lemma 4.14. Let $\eta > 0$ and $\lambda > 0$. Define $x \equiv \sqrt{r}$ and $a \equiv \sqrt{\lambda}$ and the expression $r - 2\sqrt{(1+2\eta)r\lambda/(1+\eta)} + \lambda$ can be restated as $x^2 - 2ax\sqrt{(1+2\eta)/(1+\eta)} + a^2$ with $x > 0$ and $a > 0$. The solutions to the quadratic equation $x^2 - 2ax\sqrt{(1+2\eta)/(1+\eta)} + a^2 = 0$ for x are $\left(\sqrt{1+2\eta}-\sqrt{\eta}\right)a/\sqrt{1+\eta}$ and $\left(\sqrt{1+2\eta}+\sqrt{\eta}\right)a/\sqrt{1+\eta}$. Thus if $\left(\sqrt{1+2\eta}+\sqrt{\eta}\right)a/\sqrt{1+\eta} > x > \left(\sqrt{1+2\eta}-\sqrt{\eta}\right)a/\sqrt{1+\eta}$, or equivalently, if $\left(1+3\eta+2\sqrt{\eta(1+2\eta)}\right)a/(1+\eta) > x^2 > \left(1+3\eta-2\sqrt{\eta(1+2\eta)}\right)a/(1+\eta)$, then $0 > x^2 - 2ax\sqrt{(1+2\eta)/(1+\eta)} + a^2$. Replacing x by $\sqrt{r}$ and a by $\sqrt{\lambda}$ we obtain (a). The proofs for (b) and (c) are similar. □

Proof of Lemma 4.15. Let $1 > \eta > 0$ and $\lambda > 0$, and we have $\sqrt{1+\eta} > \sqrt{1-\eta}$ and $\sqrt{1+2\eta} > \sqrt{\eta}$, therefore we have $2\sqrt{1+\eta} > \sqrt{1+\eta} +$

$\sqrt{1-\eta}$ and $2\sqrt{1+2\eta} > \sqrt{1+2\eta} + \sqrt{\eta}$. By multiplying these inequalities $4\sqrt{1+2\eta}\sqrt{1+\eta} > \left(\sqrt{1+2\eta}+\sqrt{\eta}\right)\left(\sqrt{1+\eta}+\sqrt{1-\eta}\right)$. By multiplying $\left(\sqrt{1+\eta}-\sqrt{1-\eta}\right)$ and dividing $2\eta\sqrt{1+\eta}$ on both sides we have $2\sqrt{1+2\eta}\left(\sqrt{1+\eta}-\sqrt{1-\eta}\right)/\eta > \left(\sqrt{1+2\eta}+\sqrt{\eta}\right)$. By squaring both sides $(1+2\eta)p_3 > \left(1+3\eta+2\sqrt{\eta(1+2\eta)}\right)\lambda/(1+\eta)$. □

Proof of Lemma 4.16. Note that $f(x)$ is continuous and differentiable over interval $\left[\sqrt{p_2}, \sqrt{p_3}\right]$:

$$\frac{df(x)}{dx} = -2\eta x - \sqrt{p_2}\left((1-2\eta) - \frac{r}{x^2}\right) \text{ and } \frac{d^2f(x)}{dx^2} = -2\eta - \frac{2r\sqrt{p_2}}{x^3} < 0$$

$$\left.\frac{df(x)}{dx}\right|_{x=\sqrt{p_2}} = -\sqrt{p_2} + \frac{r}{\sqrt{p_2}} = \frac{1}{\sqrt{p_2}}(r-p_2)$$

$$\left.\frac{df(x)}{dx}\right|_{x=\sqrt{p_3}} = -\sqrt{p_2} - 2\eta(\sqrt{p_3}-\sqrt{p_2}) + \frac{r\sqrt{p_2}}{p_3} = \frac{\sqrt{p_2}}{p_3}(r-r_2)$$

If $r \in (0, p_2]$, then $f(x)$ is decreasing over $\left[\sqrt{p_2}, \sqrt{p_3}\right]$, therefore $x^* = \sqrt{p_2}$. If $r \geq r_2$, then $f(x)$ is increasing over $\left[\sqrt{p_2}, \sqrt{p_3}\right]$, therefore $x^* = \sqrt{p_3}$. If $r \in (p_2, r_2)$, then $f(x)$ is increasing in the neighborhood above $x = \sqrt{p_2}$ and decreasing in the neighborhood below $x = \sqrt{p_3}$. Also note that $d^2f(x)/dx^2 < 0$, therefore $x^* \in \left(\sqrt{p_2}, \sqrt{p_3}\right)$ that satisfies the first order condition

$$\left.\frac{df(x)}{dx}\right|_{x=x^*} = 0 \Rightarrow 2\eta\left(x^*\right)^3 + (1-2\eta)\sqrt{p_2}\left(x^*\right)^2 - r\sqrt{p_2} = 0 \tag{A.1}$$

which is a cubic equation. According to the general formula for roots of cubic equation, $x^* = \sqrt{p_{cu}} \in \left(\sqrt{p_2}, \sqrt{p_3}\right)$. □

Proof of Lemma 4.17. Note that $f(x)$ is continuous and differentiable for $x \geq \sqrt{p_3}$:

$$\frac{df(x)}{dx} = -\sqrt{p_1}\left(1+2\eta-\frac{r}{x^2}\right) \text{ and } \frac{d^2f(x)}{dx^2} = -\frac{2r\sqrt{p_1}}{x^3} < 0$$

$$\left.\frac{df(x)}{dx}\right|_{x=\sqrt{p_3}} = \frac{\sqrt{p_1}}{p_3}(r-r_3) \text{ and } \lim_{x\to+\infty}\frac{df(x)}{dx} = -(1+2\eta)\sqrt{p_1} < 0$$

If $r \in (0, r_3]$, then $f(x)$ is decreasing for $x \geq \sqrt{p_3}$, therefore $x^* = \sqrt{p_3}$. If $r > r_3$, then $f(x)$ is increasing in the neighborhood above $x = \sqrt{p_3}$ and decreasing when x approaches $+\infty$. Also note that $d^2f(x)/dx^2 < 0$, therefore $x^* > \sqrt{p_3}$ that satisfies first order condition $df(x)/dx|_{x=x^*} = 0 \Rightarrow x^* = \sqrt{r/(1+2\eta)}$. □

Proof of Lemma 4.22. Let $\eta > 1/3$ and $\lambda > 0$. Note that

$$\begin{aligned} 3\eta > 1 &\Leftrightarrow 2\sqrt{\eta} > \sqrt{1+\eta} \\ &\Leftrightarrow \sqrt{1+\eta} > 2\left(\sqrt{1+\eta} - \sqrt{\eta}\right) \\ &\Leftrightarrow (1+\eta)\left(\sqrt{1+\eta} + \sqrt{\eta}\right)^2 > 4 \\ &\Leftrightarrow \left(1 + 2\eta + 2\sqrt{\eta(1+\eta)}\right)\lambda > 4\lambda/(1+\eta) \end{aligned}$$

□

Proof of Lemma 4.25. Note that $f(x)$ is continuous and differentiable for $x \geq \sqrt{p_4}$:

$$\begin{aligned} &\frac{df(x)}{dx} = -\sqrt{p_1}\left(1 + 2\eta - \frac{r}{x^2}\right) \text{ and } \frac{d^2f(x)}{dx^2} = -\frac{2r\sqrt{p_1}}{x^3} < 0 \\ &\left.\frac{df(x)}{dx}\right|_{x=\sqrt{p_4}} = \frac{\sqrt{p_1}}{p_4}(r - r_4) \text{ and } \lim_{x\to+\infty}\frac{df(x)}{dx} = -(1+2\eta)\sqrt{p_1} < 0 \end{aligned}$$

If $r \in (0, r_4]$, then $f(x)$ is decreasing for $x \geq \sqrt{p_4}$, therefore $x^* = \sqrt{p_4}$. If $r > r_4$, then $f(x)$ is increasing in the neighborhood above $x = \sqrt{p_4}$ and decreasing when x approaches $+\infty$. Note that $d^2f(x)/dx^2 < 0$, therefore $x^* > \sqrt{p_4}$ satisfies first order condition $df(x)/dx|_{x=x^*} = 0 \Rightarrow x^* = \sqrt{r/(1+2\eta)}$. □

Proof of Lemma 4.26. Let $\eta > 0$ and $\lambda > 0$, then we have

$$\begin{aligned} &\sqrt{\eta}\left(\sqrt{1+2\eta} - \sqrt{1+\eta}\right) + \eta > 0 \Leftrightarrow 0 > -\sqrt{\eta}\sqrt{1+2\eta} + \sqrt{\eta}\sqrt{1+\eta} - \eta \\ &\Leftrightarrow \sqrt{1+2\eta}\sqrt{1+\eta} > \left(\sqrt{1+2\eta} + \sqrt{\eta}\right)\left(\sqrt{1+\eta} - \sqrt{\eta}\right) \\ &\Leftrightarrow \left(\sqrt{1+\eta} + \sqrt{\eta}\right)\sqrt{1+2\eta} > \left(\sqrt{1+2\eta} + \sqrt{\eta}\right)/\sqrt{1+\eta} \end{aligned}$$

by squaring both sides we have $(1+2\eta)p_4 > \left(1 + 3\eta + 2\sqrt{\eta(1+2\eta)}\right)\lambda/(1+\eta)$.□

Proof of Lemma 5.2. Let $1 > \overline{\eta} > 0$ and $\lambda > 0$. Define $x \equiv \sqrt{p}$ and $a \equiv \sqrt{\lambda}$ and restate the expression $\overline{\eta}p/2 + 2\sqrt{(1-\overline{\eta})p\lambda} - \lambda$ as $\overline{\eta}x^2/2 + 2ax\sqrt{(1-\overline{\eta})} - a^2$ with $x > 0$ and $a > 0$. The solutions to the quadratic equation $\overline{\eta}x^2/2 + 2ax\sqrt{(1-\overline{\eta})} - a^2 = 0$ for x are $0 > -2\left(\sqrt{1-\overline{\eta}/2} + \sqrt{1-\overline{\eta}}\right)a/\overline{\eta}$ and $2\left(\sqrt{1-\overline{\eta}/2} - \sqrt{1-\overline{\eta}}\right)a/\overline{\eta} > 0$. Therefore if $2\left(\sqrt{1-\overline{\eta}/2} - \sqrt{1-\overline{\eta}}\right)a/\overline{\eta} > x > 0$, or equivalently, $4\left(\sqrt{1-\overline{\eta}/2} - \sqrt{1-\overline{\eta}}\right)^2 a^2/\overline{\eta}^2 > x^2 > 0$, then $0 >$

$\overline{\eta}x^2/2 + 2ax\sqrt{(1-\overline{\eta})} - a^2$. Replacing x by $\sqrt{p}$ and a by $\sqrt{\lambda}$ we obtain (a). The proofs for (b) and (c) are similar. □

Proof of Lemma 5.3. Let $1 > \overline{\eta} > 0$ and $\lambda > 0$, then we have

$$
\begin{aligned}
0 > \overline{\eta}^2 - 2\overline{\eta} &\Leftrightarrow \overline{\eta}^2 - 4\overline{\eta} + 4 > 2\overline{\eta}^2 - 6\overline{\eta} + 4 \\
&\Leftrightarrow (2-\overline{\eta})^2 > 4(1-\overline{\eta}/2)(1-\overline{\eta}) \\
&\Leftrightarrow 2-\overline{\eta} > 2\sqrt{1-\overline{\eta}/2}\sqrt{1-\overline{\eta}} \\
&\Leftrightarrow \overline{\eta} > 2\sqrt{1-\overline{\eta}/2}\sqrt{1-\overline{\eta}} - 2(1-\overline{\eta}) \\
&\Leftrightarrow 1/\sqrt{1-\overline{\eta}} > 2\left(\sqrt{1-\overline{\eta}/2} - \sqrt{1-\overline{\eta}}\right)/\overline{\eta} \\
&\Leftrightarrow \lambda/(1-\overline{\eta}) > 4\left(\sqrt{1-\overline{\eta}/2} - \sqrt{1-\overline{\eta}}\right)^2 \lambda/\overline{\eta}^2
\end{aligned}
$$

□

Proof of Lemma 5.4. Let $2 > \overline{\eta} > 0$ and $\lambda > 0$. Define $x \equiv \sqrt{p}$ and $a \equiv \sqrt{\lambda}$ and restate the expression $(1-\overline{\eta}/2)p - 2\sqrt{p\lambda} + \lambda$ as $(1-\overline{\eta}/2)x^2 - 2ax + a^2$ with $x > 0$ and $a > 0$. The solutions to the quadratic equation $(1-\overline{\eta}/2)x^2 - 2ax + a^2 = 0$ for x are $\sqrt{2}a/\left(\sqrt{2}+\sqrt{\overline{\eta}}\right) > 0$ and $\sqrt{2}a/\left(\sqrt{2}-\sqrt{\overline{\eta}}\right) > 0$. Therefore if $\sqrt{2}a/\left(\sqrt{2}-\sqrt{\overline{\eta}}\right) > x > \sqrt{2}a/\left(\sqrt{2}+\sqrt{\overline{\eta}}\right)$, or equivalently, $2a^2/\left(2+\overline{\eta}-2\sqrt{2\overline{\eta}}\right) > x^2 > 2a^2/\left(2+\overline{\eta}+2\sqrt{2\overline{\eta}}\right)$, then $0 > (1-\overline{\eta}/2)x^2 - 2ax + a^2$. Replacing x by $\sqrt{p}$ and a by $\sqrt{\lambda}$ we obtain (a). The proofs for (b) and (c) are similar. □

Proof of Lemma 5.5. Let $\overline{\eta} > 0$ and $\lambda > 0$, then $1 + \sqrt{2\overline{\eta}} > 0 \Leftrightarrow 2 + \sqrt{2\overline{\eta}} > 1 \Leftrightarrow \sqrt{2} > 1/\left(\sqrt{2}+\sqrt{\overline{\eta}}\right) \Leftrightarrow 4\lambda > 2\lambda/\left(2+\overline{\eta}+2\sqrt{2\overline{\eta}}\right)$. □

Proof of Lemma 5.6. Let $8/9 > \overline{\eta} > 0$ and $\lambda > 0$, then we have

$$
\begin{aligned}
8\overline{\eta} - 9\overline{\eta}^2 > 0 &\Leftrightarrow 2\sqrt{2\overline{\eta}} > 3\overline{\eta} \\
&\Leftrightarrow 2 - 2\overline{\eta} > 2 + \overline{\eta} - 2\sqrt{2\overline{\eta}} \\
&\Leftrightarrow \sqrt{2}\sqrt{1-\overline{\eta}} > \sqrt{2} - \sqrt{\overline{\eta}} \\
&\Leftrightarrow \sqrt{2}/\left(\sqrt{2}-\sqrt{\overline{\eta}}\right) > 1/\sqrt{1-\overline{\eta}} \\
&\Leftrightarrow 2\lambda/\left(2+\overline{\eta}-2\sqrt{2\overline{\eta}}\right) > \lambda/(1-\overline{\eta})
\end{aligned}
$$

and we obtain (a). The proofs for (b) and (c) are similar. □

Proof of Lemma 5.7. Let $1 > \overline{\eta} > 0$ and $\lambda > 0$. Define $x \equiv \sqrt{p}$ and $a \equiv \sqrt{\lambda}$ and restate expression $\overline{\eta}p/2 - 2\left(1 - \sqrt{1-\overline{\eta}}\right)\sqrt{p\lambda}$ as $\overline{\eta}x^2/2 - 2\left(1 - \sqrt{1-\overline{\eta}}\right)ax$ with $x > 0$ and $a > 0$. The solution to the quadratic equation $\overline{\eta}x^2/2 - 2\left(1 - \sqrt{1-\overline{\eta}}\right)ax = 0$ for x are 0 and $4\left(1 - \sqrt{1-\overline{\eta}}\right)a/\overline{\eta} > 0$. Therefore if $4\left(1 - \sqrt{1-\overline{\eta}}\right)a/\overline{\eta} > x > 0$, or equivalently, $16\left(2 - \overline{\eta} - 2\sqrt{1-\overline{\eta}}\right)a^2/\overline{\eta}^2 > x^2 > 0$, then $0 > \overline{\eta}x^2/2 - 2\left(1 - \sqrt{1-\overline{\eta}}\right)ax$. Replacing x by $\sqrt{p}$ and a by $\sqrt{\lambda}$ we obtain (a). The proofs for (b) and (c) are similar. □

Proof of Lemma 5.8. Let $8/9 > \overline{\eta} > 0$ and $\lambda > 0$, then we have

$$\begin{aligned}
0 > 9\overline{\eta}^2 - 8\overline{\eta} &\Leftrightarrow 16(1-\overline{\eta}) > 9\overline{\eta}^2 - 24\overline{\eta} + 16 \\
&\Leftrightarrow 4\sqrt{1-\overline{\eta}} > 4 - 3\overline{\eta} \\
&\Leftrightarrow 4\left(1 - \sqrt{1-\overline{\eta}}\right)/\overline{\eta} > 1/\sqrt{1-\overline{\eta}} \\
&\Leftrightarrow 16\left(2 - \overline{\eta} - 2\sqrt{1-\overline{\eta}}\right)^2 \lambda/\overline{\eta}^2 > \lambda/(1-\overline{\eta})
\end{aligned}$$

and we obtain (a). The proofs for (b) and (c) are similar. □

Proof of Lemma 5.9. Let $1 > \overline{\eta} > 0$ and $\lambda > 0$, then $1 > \sqrt{1-\overline{\eta}} \Leftrightarrow \overline{\eta} > 2\sqrt{1-\overline{\eta}} - 2 + 2\overline{\eta} \Leftrightarrow 4\lambda/(1-\overline{\eta}) > 16\left(2 - \overline{\eta} - 2\sqrt{1-\overline{\eta}}\right)\lambda/\overline{\eta}^2$. Also note that $1 > \sqrt{1-\overline{\eta}} \Leftrightarrow 2 - \overline{\eta} - 2\sqrt{1-\overline{\eta}} > 0 \Leftrightarrow 2\left(1 - \sqrt{1-\overline{\eta}}\right) > \overline{\eta} \Leftrightarrow 16\left(2 - \overline{\eta} - 2\sqrt{1-\overline{\eta}}\right)\lambda/\overline{\eta}^2 > 4\lambda$. □

Proof of Lemma 5.14. Let $8/9 > \overline{\eta} > 0$ and $\lambda > 0$. According to Lemma 5.8 and 5.9, $4\overline{p}_1 > \overline{p}_2 > \overline{p}_1$, therefore:

$$\begin{aligned}
2\sqrt{\overline{p}_1} > \sqrt{\overline{p}_2} &\Leftrightarrow \sqrt{\overline{p}_1} > \sqrt{\overline{p}_2} - \sqrt{\overline{p}_1} > 0 \\
&\Leftrightarrow 1 > \left(\sqrt{\overline{p}_2} - \sqrt{\overline{p}_1}\right)/\sqrt{\overline{p}_1} \\
&\Leftrightarrow \overline{\eta} > \overline{\eta}\left(\sqrt{\overline{p}_2} - \sqrt{\overline{p}_1}\right)/\sqrt{\overline{p}_1} \\
&\Leftrightarrow 1 > (1-\overline{\eta}) + \overline{\eta}\left(\sqrt{\overline{p}_2} - \sqrt{\overline{p}_1}\right)/\sqrt{\overline{p}_1} \\
&\Leftrightarrow \overline{p}_2 > (1-\overline{\eta})\overline{p}_2 + \overline{\eta}\overline{p}_2\left(\sqrt{\overline{p}_2} - \sqrt{\overline{p}_1}\right)/\sqrt{\overline{p}_1}
\end{aligned}$$

Since $\overline{\eta}\overline{p}_2 + (1-\overline{\eta})\sqrt{\overline{p}_1\overline{p}_2} - \overline{\eta}\overline{p}_2\sqrt{\overline{p}_2}/2\left(\sqrt{\overline{p}_2} - \sqrt{\overline{p}_1}\right) = (1-\overline{\eta})\sqrt{\overline{p}_1\overline{p}_2} + \overline{\eta}\overline{p}_2\left(\sqrt{\overline{p}_2} - \sqrt{4\overline{p}_1}\right)/2\left(\sqrt{\overline{p}_2} - \sqrt{\overline{p}_1}\right)$. Since $4\overline{p}_1 > \overline{p}_2 > \overline{p}_1$, therefore we have $(1-\overline{\eta})\overline{p}_2 > (1-\overline{\eta})\sqrt{\overline{p}_1\overline{p}_2}$ and $\overline{\eta}\overline{p}_2\left(\sqrt{\overline{p}_2} - \sqrt{\overline{p}_1}\right)/\sqrt{\overline{p}_1} > 0 >$

$\overline{\eta}\overline{p}_2\left(\sqrt{\overline{p}_2}-\sqrt{4\overline{p}_1}\right)/2\left(\sqrt{\overline{p}_2}-\sqrt{\overline{p}_1}\right)$. Thus $(1-\overline{\eta})\overline{p}_2+\overline{\eta}\overline{p}_2\left(\sqrt{\overline{p}_2}-\sqrt{\overline{p}_1}\right)/\sqrt{\overline{p}_1} > \overline{\eta}\overline{p}_2+(1-\overline{\eta})\sqrt{\overline{p}_1\overline{p}_2}-\overline{\eta}\overline{p}_2\sqrt{\overline{p}_2}/2\left(\sqrt{\overline{p}_2}-\sqrt{\overline{p}_1}\right)$. Therefore we obtain (a). According to Lemma 5.8, $\overline{p}_2 > \overline{p}_1$, therefore $(1-\overline{\eta})(\overline{p}_2-\overline{p}_1)+\overline{\eta}\overline{p}_2\left(\sqrt{\overline{p}_2}-\sqrt{\overline{p}_1}\right)/\sqrt{\overline{p}_1} > 0 \Leftrightarrow (1-\overline{\eta})\overline{p}_2+\overline{\eta}\overline{p}_2\left(\sqrt{\overline{p}_2}-\sqrt{\overline{p}_1}\right)/\sqrt{\overline{p}_1} > (1-\overline{\eta})\overline{p}_1 = \lambda$. Therefore we obtain (b). □

Proof of Lemma 5.15. Note that $f(x)$ is continuous and differentiable over interval $\left[\sqrt{\overline{p}_1},\sqrt{\overline{p}_2}\right]$:

$$\frac{df(x)}{dx} = -\overline{\eta}x-\sqrt{\overline{p}_1}\left((1-2\overline{\eta})-\frac{r}{x^2}\right) \text{ and } \frac{d^2f(x)}{dx^2} = -\overline{\eta}-\frac{2r\sqrt{\overline{p}_1}}{x^3} < 0$$

$$\left.\frac{df(x)}{dx}\right|_{x=\sqrt{\overline{p}_1}} = -(1-\overline{\eta})\sqrt{\overline{p}_1}+\frac{r}{\sqrt{\overline{p}_1}} = \frac{1}{\sqrt{\overline{p}_1}}(r-\lambda)$$

$$\left.\frac{df(x)}{dx}\right|_{x=\sqrt{\overline{p}_2}} = -\overline{\eta}\sqrt{\overline{p}_2}-(1-2\overline{\eta})\sqrt{\overline{p}_1}+\frac{r\sqrt{\overline{p}_1}}{\overline{p}_2} = \frac{\sqrt{\overline{p}_1}}{\overline{p}_2}(r-\overline{r}_2)$$

If $r \in (0,\lambda]$, then $f(x)$ is decreasing over $\left[\sqrt{\overline{p}_1},\sqrt{\overline{p}_2}\right]$, therefore $x^* = \sqrt{\overline{p}_1}$. If $r \geq \overline{r}_2$, then $f(x)$ is increasing over $\left[\sqrt{\overline{p}_1},\sqrt{\overline{p}_2}\right]$, therefore $x^* = \sqrt{\overline{p}_2}$. If $r \in (\lambda,\overline{r}_2)$, then $f(x)$ is increasing in the neighborhood above $x = \sqrt{\overline{p}_1}$ and is decreasing in the neighborhood below $x = \sqrt{\overline{p}_2}$. Also note that $d^2f(x)/dx^2 < 0$, therefore $x^* \in \left(\sqrt{\overline{p}_1},\sqrt{\overline{p}_2}\right)$ that satisfies the first order condition

$$\left.\frac{df(x)}{dx}\right|_{x=x^*} = 0 \Rightarrow \overline{\eta}\left(x^*\right)^3+(1-2\overline{\eta})\sqrt{\overline{p}_1}\left(x^*\right)^2-r\sqrt{\overline{p}_1} = 0 \qquad \text{(A.2)}$$

which is a cubic equation. According to the general formula for roots of cubic equation, $x^* = \sqrt{\overline{p}_{cu}} \in \left(\sqrt{\overline{p}_1},\sqrt{\overline{p}_2}\right)$. □

Proof of Lemma 5.16. Note that $f(x)$ is continuous and differentiable for $x \geq \sqrt{\overline{p}_2}$:

$$\frac{df(x)}{dx} = -\sqrt{\lambda}\left(1-\frac{r}{x^2}\right) \text{ and } \frac{d^2f(x)}{dx^2} = -\frac{2r\sqrt{\lambda}}{x^3} < 0$$

$$\left.\frac{df(x)}{dx}\right|_{x=\sqrt{\overline{p}_2}} = \frac{\sqrt{\lambda}}{\overline{p}_2}(r-\overline{p}_2) \text{ and } \left.\frac{df(x)}{dx}\right|_{x\to+\infty} = -\sqrt{\lambda} < 0$$

If $r \in (0,\overline{p}_2]$, then $f(x)$ is decreasing for $r \geq \sqrt{\overline{p}_2}$, therefore $x^* = \sqrt{\overline{p}_2}$. If $r > \sqrt{\overline{p}_2}$, then $f(x)$ is increasing in the neighborhood above $x = \sqrt{\overline{p}_2}$. Since $f(x)$ is decreasing as $x \to +\infty$ and since $f(x)$ is concave $(d^2f(x)/dx^2 < 0)$, therefore $x^* > \sqrt{\overline{p}_2}$ is solved from first order condition $df(x)/dx|_{x=x^*} = 0 \Rightarrow x^* = \sqrt{r}$. □

Proof of Lemma 5.18. Let $2 > \overline{\eta} \geq 8/9$ and $\lambda > 0$, then we have

$$\begin{aligned} 2\overline{\eta} \geq 16/9 &\Leftrightarrow \sqrt{2\overline{\eta}} \geq 4/3 > 1 \\ &\Leftrightarrow 1 > 2 - \sqrt{2\overline{\eta}} \\ &\Leftrightarrow \sqrt{2}/\left(\sqrt{2} - \sqrt{\overline{\eta}}\right) > 2 \\ &\Leftrightarrow 2\lambda/\left(2 + \overline{\eta} - 2\sqrt{2\overline{\eta}}\right) > 4\lambda \end{aligned}$$

□

Proof of Lemma 5.21. Note that $f(x)$ is continuous and differentiable for $x \geq \sqrt{\overline{p}_3}$:

$$\frac{df(x)}{dx} = -\sqrt{\lambda}\left(1 - \frac{r}{x^2}\right) \text{ and } \frac{d^2f(x)}{dx^2} = -\frac{2r\sqrt{\lambda}}{x^3} < 0$$

$$\left.\frac{df(x)}{dx}\right|_{x=\sqrt{\overline{p}_3}} = \frac{\sqrt{\lambda}}{\overline{p}_3}(r - \overline{p}_3) \text{ and } \left.\frac{df(x)}{dx}\right|_{x\to+\infty} = -\sqrt{\lambda} < 0$$

If $r \in (0, \overline{p}_3]$, then $f(x)$ is decreasing for $x \geq \sqrt{\overline{p}_3}$, therefore $x^* = \sqrt{\overline{p}_3}$. If $r > \sqrt{\overline{p}_3}$, then $f(x)$ is increasing in the neighborhood above $x = \sqrt{\overline{p}_3}$. Since $f(x)$ is decreasing as $x \to +\infty$ and since $f(x)$ is concave $(d^2f(x)/dx^2 < 0)$, therefore $x^* > \sqrt{\overline{p}_3}$ is solved from first order condition $df(x)/dx|_{x=x^*} = 0 \Rightarrow x^* = \sqrt{r}$. □

Proof of Lemma 5.23. Let $\overline{\eta} > 2$ and $\lambda > 0$. Define $x \equiv \sqrt{p}$ and $a \equiv \sqrt{\lambda}$ and restate the expression $(1 - \overline{\eta}/2)p - 2\sqrt{p\lambda} + \lambda$ as $(1 - \overline{\eta}/2)x^2 - 2ax + a^2$ with $x > 0$ and $a > 0$. The solutions to the quadratic equation $(1 - \overline{\eta}/2)x^2 - 2ax + a^2 = 0$ for x are $\sqrt{2}a/\left(\sqrt{2} + \sqrt{\overline{\eta}}\right) > 0$ and $0 > \sqrt{2}a/\left(\sqrt{2} - \sqrt{\overline{\eta}}\right)$. Therefore if $\sqrt{2}a/\left(\sqrt{2} + \sqrt{\overline{\eta}}\right) > x > 0$, or equivalently, $2a^2/\left(2 + \overline{\eta} + 2\sqrt{2\overline{\eta}}\right) > x^2 > 0$, then $(1 - \overline{\eta}/2)x^2 - 2ax + a^2 > 0$. Replacing x by $\sqrt{p}$ and a by $\sqrt{\lambda}$ we obtain (a). The proofs for (b) and (c) are similar. □

References

Artzner, P., Delbaen, F., Eber, J.-M., Heath, D.: Coherent measures of risk. Math. Financ. **9**(3), 203–228 (1999)

Bolton, P., Dewatripont, M.: Contract Theory. MIT Press, Cambridge (2005)

Filiz-Ozbay, E., Guryan, J., Hyndman, K., Kearney, M.S., Ozbay, E.Y.: Do lottery payments induce savings behavior: evidence from the lab. Technical report, National Bureau of Economic Research (2013)

Fudenberg, D., Tirole, J.: Moral hazard and renegotiation in agency contracts. Econom.: J. Econom. Soc. **58**, 1279–1319 (1990)

Fukunaga, K., Huffman, W.E.: The role of risk and transaction costs in contract design: evidence from farmland lease contracts in us agriculture. Am. J. Agric. Econ. **91**(1), 237–249 (2009)

Grossman, S.J., Hart, O.D.: An analysis of the principal-agent problem. Econom.: J. Econom. Soc. **51**, 7–45 (1983)

Guesnerie, R., Laffont, J.-J.: A complete solution to a class of principal-agent problems with an application to the control of a self-managed firm. J. Public Econ. **25**(3), 329–369 (1984)

Harris, M., Raviv, A.: Some results on incentive contracts with applications to education and employment, health insurance, and law enforcement. Am. Econ. Rev. **68**, 20–30 (1978)

Hölmstrom, B.: Moral hazard and observability. Bell J. Econ. **10**, 74–91 (1979)

Hölmstrom, B., Milgrom, P.: Aggregation and linearity in the provision of intertemporal incentives. Econom.: J. Econom. Soc. **55**, 303–328 (1987)

Kahneman, D., Tversky, A.: Prospect theory: an analysis of decision under risk. Econom.: J. Econom. Soc. **47**, 263–291 (1979)

Kim, S.-H., Cohen, M.A., Netessine, S., Veeraraghavan, S.: Contracting for infrequent restoration and recovery of mission-critical systems. Manag. Sci. **56**(9), 1551–1567 (2010)

Levy, H.: Stochastic dominance and expected utility: survey and analysis. Manag. Sci. **38**(4), 555–593 (1992)

Lewis, G., Bajari, P.: Moral hazard, incentive contracts, and risk: evidence from procurement. Rev. Econ. Stud. rdu002 **81** (2014)

Luenberger, D.G.: Microeconomic Theory, vol. 486. McGraw-Hill, New York (1995)

March, J.G., Shapira, Z.: Managerial perspectives on risk and risk taking. Manag. Sci. **33**(11), 1404–1418 (1987)

Maskin, E., Tirole, J.: The principal-agent relationship with an informed principal: the case of private values. Econom.: J. Econom. Soc. **58**, 379–409 (1990)

Maskin, E., Tirole, J.: The principal-agent relationship with an informed principal, ii: common values. Econom.: J. Econom. Soc. **60**, 1–42 (1992)

S. Zeng, M. Dror, *Formulating Principal-Agent Service Contracts for a Revenue Generating Unit*, SpringerBriefs in Operations Management,
DOI 10.1007/978-3-319-18672-6

Myerson, R.B.: Mechanism design by an informed principal. Econom.: J. Econom. Soc. **51**, 1767–1797 (1983)

Pratt, J.W.: Risk aversion in the small and in the large. Econom.: J. Econom. Soc. **32**, 122–136 (1964)

Ross, S.A.: The economic theory of agency: the principal's problem. Am. Econ. Rev. **63**(2), 134–139 (1973)

Ross, S.M.: Introduction to Probability Models (2006). Access Online via Elsevier

Rothschild, M., Stiglitz, J.E.: Increasing risk: I. a definition. J. Econ. Theory **2**(3), 225–243 (1970)

Stiglitz, J.E.: Incentives and risk sharing in sharecropping. Rev. Econ. Stud. **41**, 219–255 (1974)

Stiglitz, J.E.: Equilibrium in product markets with imperfect information. Am. Econ. Rev. **69**(2), 339–345 (1979)

Tversky, A., Kahneman, D.: Advances in prospect theory: cumulative representation of uncertainty. J. Risk Uncertain. **5**(4), 297–323 (1992)

Index

S. Zeng, M. Dror, *Formulating Principal-Agent Service Contracts for a Revenue Generating Unit*, SpringerBriefs in Operations Management,
DOI 10.1007/978-3-319-18672-6

Kolibri 360 GmbH
Ferdinand-Jühlke-Straße 7
99095 Erfurt, Deutschland
produktsicherheit@kolibri360.de